Moral Habitat

SUNY series on Religion and the Environment

Harold Coward, editor

Moral Habitat

Ethos and Agency for the Sake of Earth

Nancie Erhard

State University of New York Press

Published by
State University of New York Press, Albany

For information, contact State University of New York Press, Albany, NY
www.sunypress.edu

Production by Ryan Morris
Marketing by Michael Campochiaro

Library of Congress Cataloging-in-Publication Data

Erhard, Nancie, 1957–
    Moral habitat : ethos and agency for the sake of earth / Nancie Erhard.
        p. cm. — (SUNY series on religion and the environment)
    Includes bibliographical references and index.
    ISBN-13: 978-0-7914-7141-8 (hardcover : alk. paper)
 1. Human ecology—Moral and ethical aspects.   2. Environmental ethics.   I. Title.

GF80.E74   2007
179'.1—dc22   2006027536

                              10 9 8 7 6 5 4 3 2 1

Contents

Acknowledgments

I could not have wished for better mentors and companions on this journey. My special thanks go to Larry Rasmussen, for his gentle guidance and inspiration, and to my life partner Mauritz, for his love, generosity, and steadfast belief in me and this work. I would also like to thank William P. Brown, Daniel T. Spencer, and emillie m. townes, as well Nancy Berlinger, Cynthia Moe-Lobeda, and David Wellman. I owe more than I can repay to Isabelle Knockwood for her hospitality and honesty. Joyce Rillett-Wood and Alan Cooper helped me fall in love with the Hebrew Bible. Marilyn Legge sparked my questions about moral agency and gave helpful advice. Paul Bowlby, Anne-Marie Dalton, Joe Foy, and Fred Krieger were always there to keep me going when things got tough, as they did several times. I'd like to extend a special thanks to my mother, brother, and sisters, especially Susan, for their understanding when I couldn't be there for them. Finally, to all the living beings at Red Brook Farm—particularly Pangur Ban, Tucker, Chaia, and Max—this is for you.

Introduction

In the early part of the twentieth century, George Santayana reported a conversation he had with "a Californian" who told him that if the philosophers had lived among the California mountains, "their systems would have been different from what they are. Certainly very different from what those systems are which the European genteel tradition has handed down since Socrates."[1] For anyone who has stood in the deep green twilight of a grove of giant redwood trees, his argument is not hard to imagine. In Plato's dialogue *Phaedrus*,[2] Socrates declares that the trees and fields beyond the city walls of Athens have nothing to say to him. His teachers are men (and he means, indeed, men); his place is in the city streets. This was a philosophic endeavor bent on the discovery of universal truths and on transcendence of the vagaries of the physical world. In the presence of the redwoods, where a human being at full height stands nestled in the protruding roots of such a massive living creature, could a philosophy so single-mindedly concentrate itself on human thought and stature? Could anyone take seriously the idea that plants and animals were made for the sake of human beings? This is exactly the difference the mountains and trees would have made, according to this anonymous Californian, who characterizes the systems of "the philosophers" as "egotistical; directly or indirectly they are anthropocentric, and inspired by the conceited notion that man (*sic*), or human reason, or the human distinction between good and evil, is the center and pivot of the universe."[3] The mountains and woods themselves, he said, "should make you at last ashamed to assert" such notions.

That a geographic location, with its topography and forms of life, could make a difference to ethics is one idea I will explore in this work. It intertwines with another: that such things as mountains and woods can, do, and should exert a morally formative influence (shame in the reported

1

conversation above), which accords a form of agency—can we call it moral agency?—to other-than-humankind. I am drawn to these ideas because, as I will later explain, the constructions of moral agency prevalent in Western ethics are inadequate to the enormous, unchosen task we have been given: healing Earth.

I'm assuming here, of course, that the peril is real, that there is at present an increasing impairment of the potential of the miraculous life of this planet to sustain itself in any way that we would recognize as flourishing, free, or wild. Even if we cannot say for certain what the consequences of our ways of life are, the risks are simply too great to continue as we are until we have certainty. Given the magnitude of the need for change, it is necessary to approach what it means to live in a moral universe and how we understand moral agency in radically different ways. I began this work by asking these questions: What would a more adequate account of moral agency be like? According to whom? What would it mean for our understanding of moral agency if we entertained the idea that moral agency was widely distributed in the biotic community, and that ethical norms arise from within *that* community? How then would we speak of human moral agency?

I want to explore for ethics what it means if we think of humans, like all other creatures, as living in habitats, habitats which for humans (though not exclusively for humans)[4] are necessarily natural-cultural. By employing the term "nature-culture"[5] in this way, I mean to emphasize that cultures do not sit on top of their biotic community, or apart from it, but are created and sustained within and by it. This includes the "subtle web of values, meanings, purposes, expectations, obligations, and legitimations that constitutes the operating norms of a culture," as Max Stackhouse describes *ethos*.[6] The development of the concept of ethos as moral habitat is intended to provide a holistic approach to ethics in which we may discover ways of thinking about our world and ourselves that will help us live more justly, sustainably, and creatively together. By "holistic" I mean that it includes not only human thought, feelings, bodies, social structures and constructs, but also the whole Community of Life as subjects, not objects. Ethos as moral habitat honors the diversity of nature-cultures and religious traditions without absolute relativism. In this work, I outline the concept of ethos as moral habitat and explore its implications in a discussion of moral agency.

The choice to discuss moral agency is not an arbitrary one. According to William P. Brown, an ethos can be conceived as a cultural, moral sphere analogous to the biosphere—in other words, a moral habitat—where human ethics and human moral agency is formed and enabled.[7] I use the phrase "moral habitat" not only to name this sphere but to emphasize that it is part of and *shaped in part by* a larger biotic community. Not only human agency, but other-than-human agency, and the relation between them, therefore, intrinsically come into play in the development of ethos as "moral habitat." With this conceptual and imaginative groundwork in place, I can explore two questions in a sequence reversed from common Western philosophical practice: In what ways can we talk about the biotic community as having moral import and agency? And *then*—what does this mean for our understanding of human moral agency?

Why this order? Santayana's anonymous Californian has certainly not been alone in calling for a nonanthropocentric philosophy and ethics. But there are acknowledged and unacknowledged problems in attempting to construct nonanthropocentric ethics by simply extending to the rest of the biosphere ethical methods, categories, and ideas—rights, for example, or sentience, subjectivity, moral considerability, or "standing"—that have been developed within an anthropocentric tradition. And although Western ecological ethics may proceed from a variety of starting points, even those most critical of anthropocentrism usually adopt these categories unreflectively. What I am saying is that the categories themselves may need to shape-shift, different concepts and methods may need to be developed, and less apparent structures of thought come into view. This is what I am up to with the category of moral agency. I am deliberately not "extending" it. Rather, I am asking how we might reconceive it in a less anthropocentric, androcentric, and ethnocentric context.

One of the less-acknowledged problems of rooting ethics in Earth by "extending" existing ethical notions is the eurocentric nature of prevailing ethical practices in the European and North American academy and church. Certain notions about thought and personhood go unquestioned, and important information may get obscured in a universalist approach to understanding what is real, true, or valued. While universality is at times a useful idea for pursuing justice and right relations, as with the UN Universal Declaration of Human Rights in the mid-twentieth century, it is not helpful or accurate to universalize offhandedly "the" human and "culture" as if all human activity and situations were uniform. This is particularly objectionable when the responsibility for and impact of ecological havoc is uniformly assigned to all humankind, because race, class, gender, and geographic location often make a great deal of difference on both counts.[8] Perhaps more subtle, but no less problematic, is that even when indigenous cultures are held up approvingly as exemplary in ecological ethics, there is frequently an assumption nevertheless of the normativity of European and Euro-American cultural categories and epistemologies, or of the nature and role of human reason. This means judgment of native peoples' lifeways according to the criteria of an alien folk.[9]

Moreover, if the most basic understanding of what may constitute moral agency involves knowing the good as well as being able to do it, then certain questions about what it means to know, who can know, and how, may be foreclosed by such an assumption.

That is why I take a cross-cultural approach to my questions, with illustrations from three different natural-cultural contexts: indigenous Northeast American (Algonquian, primarily Mi'kmaq) traditions, the Hebrew Bible, and post-Cartesian science. The exploration of these diverse nature-cultures serves two interrelated purposes. One is to acknowledge the role of the whole biotic community in the formation of an ethos. The second is to provide illustrations of how moral agency is constructed in a variety of terms other than those familiar to Western modernity. It is working from this basis—the concept of ethos as moral habitat and a cross-cultural approach to moral agency—that I propose a reconsideration of theories of human moral agency. Finally, because the task of ethics is prescriptive as well as descriptive, I show how ethos as "moral habitat" can be a productive working concept for considering the transformation of human agency on behalf of the flourishing of the whole Earth community and how certain practices can weave a morally responsible ethos for the flourishing of Earth.

My interest in the participation of the biotic community in moral formation and in the ways we understand the moral life of otherkind arises out of my childhood experience of what David Abram calls the "more-than-human world,"[10] out of my adult experiences as an immigrant, and through a long-term process of becoming acquainted with the traditions and situations of North American First Nations. This study originated in grief and loneliness—and deep joy. The human social world in which I was raised told me that the rest of the natural world was not only wholly Other, it was "dumb" matter in a dead universe.[11] That was simply not the way I found it, something I'm sure is common among children who grow up, as I did, with some acquaintance with the more-than-human world. Other animals, insects, winds, stars, and plants, especially trees, told me stories about themselves and about myself and God quite different from what I heard in school, from Disney, or, for the most part, at home and in church. I could not have articulated it then, but their company had the quality of holy friendship. I became a different person than I would have been without it. While I had that company, I often felt lonely when the human world denied or trivialized that friendship. Nevertheless, I felt comforted and nurtured by this other-than-human friendship many ways. I found myself truly bereft when I grew older and that other-than-human world went silent around me. It was no longer the full silence of companionship; an empty silence, one of absence and loss, took its place.

The silence was partially a cultural deadening of perception finally setting in. The power of an ethos is difficult to resist continually. But it was also an actual absence. Our working-class subdivision was one of the first ripples of what became a tidal wave of postwar sprawl that has yet to abate. During my childhood, next to my elementary school, the last market farm for miles was cultivated with the aid of the muscle, bone, brainpower, and excrement of two huge mules. The creek that ran between farm and school was no wilderness, even in a child's sturdy imagination, but nevertheless it used to buzz with myriad insects, chirp with crickets, and thrum, croak, and sing with frogs. A couple of years after I went on to high school, the bulldozers came. An apartment house complex was built (uninsulated, in the teeth of the early seventies' oil embargo) on the site of the farm. Fragrant tomato plants and soil the color of dark chocolate, fed by mule droppings for decades, were smothered with black tar asphalt that softened and stank under the summer sun; the creek took on an oily sheen and went silent. The roar of air conditioners and the highway overwhelmed all other sound. Something of me was also smothered and silenced, and I was painfully aware of it.

Perhaps it is the coincidence of place and time and my growing-up self that the cultural deafness around me and the destruction of life in front of me became so clearly connected in my mind, with each other and with despair. My father drinking himself to sleep during the nightly news with its pictures of dogs and fire hoses turned on peaceful marching people, pictures of assassinations, napalm bombs, burning forests, and screaming naked children; the fumes from the chemical factory (Monsanto) a couple of miles away; the roar of the highway and stink of asphalt—all these were connected with the idea of dumb matter in a dead universe. Something was terribly wrong, and there seemed little I or anyone else could do about it. Like many young people at that time, I set out in search of alternative ways to live. For me, this was not a youthful lark but a matter of survival. It was an intellectual, physical, emotional, and spiritual search, with many detours and dangers. In a way too convoluted to tell, my search led to formal study and, finally, to living communities.

Gradually, I have begun to feel restored to a community such as the one I knew in my childhood; albeit quite different, it is one that is not only human. This is the source of tremendous joy. I am home. Physically and culturally, my home is very distant from where I was raised. It is in another country, in a place with a quite different ethos from the suburban midwest United States. I now live on the fringe of Earth's largest tidal zone. Depending upon the dance of earth, sun, and moon, a muddy tide trickles in or flows between or swells clear over the banks of the brook below our house to form a pond. It settles and then ebbs. Eagles and great blue herons appear with the movement of the tide. In spring, gasperaux (a kind of shad) struggle up from the sea to the stream to the beaver pond at its head. This is also a land shaped by human hands. The old growth forests that once covered the uplands are long gone, cut for ship timber in the days of sail. And the Acadians had built dikes to form fields rich with silt deposited by the tides, flushed of salt by rainwater. But even with the dikes, the expression "coast line" has no meaning here; there is no line, only constant flux. Living where the salt water meets the sweet, beside dikeland and salt marshes that give way to glistening mud flats, I share a space at once marginal and wide with myriad creatures—land, sea, and intertidal—for neighbors.

Added to this other-than-human community is that, among a variety of human beings in various natural-cultural communities, I have found kindred spirits. Again, I would not have become the same person had I not come here, to the land bounded, and in many ways culturally defined, by the Atlantic Ocean, the Northumberland Strait, and the Bay of Fundy. Out of my experience here my conviction arises that place matters to moral formation and so to ethics. And my commitment to this place has led me to study these particular nature-cultures, which are highly influential in it. They are also of importance in the whole North American context. While there are other equally important nature-cultures in both contexts, the scope of this project called for some limitation.

Since I immigrated to Mi'kmaki[12] in 1984, and especially since 1989, I have had the gift of friendship and conversations, generously shared stories and ceremonies, with Mi'kmaq people, especially Isabelle Knockwood.[13] The differences in the constructions of otherkind between traditional Mi'kmaw and Western thought became self-evident, even to a non-Native like me. They were apparent to me before I began any formal study of ethics or had even heard the term "moral agency." I will deal with those differences and the misperceptions of native thought perpetuated by Euro-American caricatures and co-optation later. What I want to note here is that it was in trying to describe the difference that I discovered the eurocentric category of moral agency, not vice versa.

I am not Mi'kmaw and do not claim to speak for them. I am not a full insider in any of the worlds investigated for this study. I honestly don't know of one where I would be, unless it is defined very broadly—educated Christian Anglophone North American. For me, within each of those descriptive elements the degree of variation from the norm is fairly high, as a dual citizen, for instance, in a different kind of intertidal zone. Living a somewhat intensified practice of cultural translation as an immigrant in a multicultural society has heightened my interest in and sensitivity to natural-cultural distinctiveness. The best I can do is bring this practice to bear, along with my formal education, and hold my understandings accountable to those who are better placed in these natural-cultural worlds than I.

Not only was Mi'kmaw nature-culture the source of the question about moral agency, it is the nature-culture with longest tenure in this place. In the scheme of ethos as moral habitat, it would then have the longest continuous shaping by the biotic community of this location. As such it is accorded a place of precedence. By placing it first in the discussion of moral agency of otherkind, I am trying to avoid presenting eurocentric ideas as generic, and then indigenous ones as alternative. Its placement does not imply that this nature-culture has been superseded by the influence of the ones following it, though they have had tremendous impact on the Mi'kmaq.

I have included the Hebrew Bible for several reasons. Not only do various ethnic and racial groups hold it as scripture, but various faith traditions share it as well. I do it to counter facile readings of it by some members of those faith traditions and by environmentalists, readings that make it appear that scripture gives license to humankind to do as they please with the rest of creation, or that it singles out humankind in a way that it does not, namely, as the only creature in intimate relation with and capable of responding to God. Many others are working on this same objective, and I don't claim that this project is more than a tiny thrust toward that end. It is also an attempt at drawing on our own wells, listening to the voices of those we claim as ancestors.

Even as we claim these texts as those of our ancestors and forebears, their interpretation is already a cross-cultural process. Those who could trace genetic linkage and live in the same geographic locations no longer live in the communities that composed them originally. But the life of these texts exists not as crystalizations of a particular moment in time and place, but in the living traditions, from the Latin, *tradere*, to hand on, in which they speak and have meaning, in many moral habitats distant from their origins. And as Carol A. Newsom so wryly observes, "Nobody owns the Bible. Though from time to time various religious traditions have attempted to restrict who might read and interpret it, the Bible always manages to evade its chaperones and sneak out for a tryst with unauthorized interpreters."[14]

In doing this kind of cross-cultural ethics there is a fine line between honoring the wisdom of people other than one's own and misappropriating their culture. As Isabelle Knockwood said to me years ago when we were first acquainted, white people have taken it all, the land and its bounty. "All we have left is our spirituality. This is not yours to take." She said she was open to speaking with me at that time because I was interested in reexamining my own cultural traditions for their neglected Earth wisdom, and because I was not another white "wannabe Indian." (At least, by the time I met Isabelle in 1990, I wasn't. Anyone who knew me as an adolescent knows I went through that, too.) I thought then, and even more so now, that part of that process of reclamation of ancestral voices was listening to the voices of those who have been harmed by the dominant and dominating patterns of Euro-North American cultures. It gives us new ears for what we may have heard all our lives.

I have included science because of its dominance in North American consciousness as a source of reliable knowledge, the historical role it has played, and its present mythic power. As Mary Midgley says, "Any system of thought playing the huge part that science plays in our lives must also shape our guiding myths and colour our imaginations profoundly. It is not just a useful tool. It is also a pattern that we follow at a deep level in trying to meet our imaginative needs."[15] By post-Cartesian science I mean to indicate those pursuits within science that have at least to some degree managed to discard the powerful myths of Descartes, his dualistic construction of mind and body, his mechanistic conceptions of life (the extreme being the truly bizarre idea that animals are literally machines without mind or feeling), and his approach to epistemology.[16] For the purposes of this study, I am drawing on research in evolution and ethology. By no means does post-Cartesian science represent the mainstream of the scientific community, however. I realized recently how potent Descartes still is in the scientific imagination when I heard a prominent biologist refer to micro-organisms as "Nature's nano-machines."

Science is so powerful a force in our context that it is almost habitual to use it as an ultimate arbiter of what is so. We often overlook the ways in which science self-limits what it can know and mistakenly treat its knowledge as total. I work from the premise that adequacy of knowledge cannot be determined solely by the limited terms that science has set for itself, as useful as they may be for specific purposes. Descartes tried to find solid ground for what could be known by eliminating the means of knowing to some bare, ultimately reliable minimum. Following along the lines of Descartes, modernity and the science that arose as an integral part of it presuppose that restriction of the means and sources of knowledge increases the reliability of that knowledge. Although there were oppositional ideas within modernity as to which means of obtaining knowledge was most reliable (i.e., whether by manipulation and measurement of the physical or by "reason"), this trait is pervasive. What if we turned this impetus around 180 degrees? What if we searched for as many ways to know as much as possible, recognizing that all knowledge is indeed "situated knowledge"?[17] We might discover that the minimalist approach that gave us "dumb matter in a dead universe" has made us unlearn some vitally important things. The regard for multiple sources of truth is a common characteristic of indigenous American thought and is one way in which it thoroughly informs my work.

The primary difference is that our basis of knowledge would attempt to be as wide and deep as possible, rather than as narrow. This study is an attempt to operate with this approach to knowledge and reality, rather than a philosophical argument for such a theory. This is not a straightforward and simple process, falling into a predictable pattern in which criteria can be weighted or arranged in a hierarchy, or are even consistently valid, in obtaining reliable knowledge and judging knowledge-organizing interpretations. And from an ethical standpoint these are also not the only criteria; there is also a consideration of consequences— social, material, and spiritual.

I recognize that the three natural-cultural worlds treated here are not discrete, but interact and overlap. To borrow Eric Wolff's apropos image, cultures are not solid objects that bounce off one another like billiard balls.[18] The Hebrew Bible and post-Cartesian science are both shared in some ways by the Mi'kmaq. Many Mi'kmaq, for example, are Christians, and while the Hebrew Bible has been a source of colonialist myths used against them, their faith is nevertheless shaped both by the Hebrew Bible and their Euro-Canadian neighbors. Both Mi'kmaq and Euro-Canadian communities also negotiate a world commonly framed by science.

The overlap and permeable boundaries among the natural-cultural worlds I explore demonstrate the aptness of the metaphor of habitat, which also has permeable boundaries. They pose no insurmountable difficulty in the pursuit of the discussion of moral agency, since my approach to these three natural-cultural habitats is not to compare and contrast them or to arbitrate among them, as if only one version of the world were "true" or any of them need be "pure." I am not holding any one of these three worlds up as a model of ecological integrity to be imitated. There are elements of sound ecological wisdom in all of them, but except for post-Cartesian science, none were constelled in the context of an ecological crisis of the magnitude we face. Nevertheless, in their own way each of them faced ecological limitations and smaller crises that were formative of their ethos.

Difficulties remain in an inductive exploration of how "agency" is conceived cross-culturally. While every culture defines for itself in some way what is normative or "moral," "agency" is a category of Western philosophy. What I am attempting to do here is to break down the construct of "moral agency" as it is used in modern Western philosophy back into its components, to challenge the implicit claims of what it means to know the good, as well as to do it. I define "moral agency" loosely, broadly, as the capacity to know the good and act on it, and will hold this definition subject to insights coming from the alternate approaches and concepts within the natural-cultural worlds we investigate. I look to see how other-than-human capacity to know and act is conceived, whether other-than-human behavior is understood to have moral content, how human moral imagination has been shaped by other-than-humankind, and where these intersect.

In the three "worlds" that I treat, I am pursuing illustrations, rather than a systematic, comprehensive, or exhaustive investigation. Because the Bible is not the primary focus of this work, for example, I have selected one biblical strand, the Priestly Creation-Flood narrative, to examine in some depth, and suggest other texts for further pursuit of the question. This illustrative approach holds for the other natural-cultural traditions as well. I follow the pattern of Yi-Fu Tuan in drawing on the work of others and on prototypical instances to make general observations.[19] My purpose is to demonstrate how the conceptual apparatus of ethos as moral habitat can bring differing ways of understanding other-than-human agency into conversation, not to make a comprehensive study of agency in each world.

Finally, while I will explain the various challenges Earth is presenting to the prevailing Western theories of agency, I am not attempting a definitive reconstruction of moral agency theory that will address all of these difficulties. I raise these questions to uproot some too-comfortable assumptions. I want to unveil the ways in which certain notions of personal responsibility inherent in agency theories contribute to the evasion of responsibility for collective, systemic, historical, or future harm.

Chapter One

Ethos as Moral Habitat

Ethos: Web of Moral Life

Max Stackhouse said that one of the distinctive tasks of ethics is "to define the ethos, that is, to identify, to evaluate, to arrange or rearrange those networks of norms that obtain in a sociocultural setting."[1] If that is so, then it belongs to ethics to consider not just a given ethos but also what makes up ethos itself, how in general networks of norms form and change, and how they function in the moral life—particularly if we are interested in attempting to arrange or rearrange such a network. I begin this task by introducing the concept of "moral habitat" as metaphor for ethos, in order to consider the ways in which the rest of the natural world participates in the formation and functioning of an ethos.

By introducing "moral habitat" as a metaphor, I want to draw attention to the role that metaphor itself plays in the process I attempt to describe. We constantly use metaphors that build on our sensory experiences to describe our subjective experiences, whether we are aware of it or not. We move from the literal meaning of a word like "grasp" to the metaphorical meaning of comprehension when we change the predicate from something we can hold with our hands, like a ball or stick, to something we "take hold of" with our minds. Metaphors are a way of mapping across conceptual domains, and they structure the concepts we use for moral reasoning at a deep level. Similarly, ethos as moral habitat is metaphorical when it is analogous, mapping the subjective experience of living in communities of norms, meanings, values, and feelings by reference to the experience of living in biotic habitats that provide the substance for and limits to our physical lives. To use the analytical terms of Lakoff and Johnson, the source domain is the biotic habitat and the target domain is the ways in which we live within community-generated norms and meanings that form, deform, and enable us to perform as

11

moral agents. We already use the language of moral habitat metaphorically when we talk about "setting behavioral boundaries" or "enlarging our horizons." I will be using the phrase "moral habitat" in one sense that is explicitly metaphorical, but I also dare, on another level, to take it as a literal expression as well, in which the word "habitat" in its conventional biotic meaning is simply modified by the adjective "moral."

The metaphorical meaning of ethos as moral habitat is not imposed arbitrarily but grows "organically," if you will, out of the ancient usage of the word "ethos." While Aristotle used it to mean moral character, and it has entered Western philosophy with that meaning, Homer used the word (in the plural) to refer to the homes or accustomed places of animals in both the *Iliad* and the *Odyssey*.[2] This is not an eccentricity of Homer. Herodotus similarly applied it to the habitual places of lions, and Oppianus to those of fish.[3] Paul Lehmann finds it "humiliating, but . . . instructive to recall that the term was first applied not to humans, but to animals."[4] He translates the ancient Greek ethos (ηθος) with the words "stall" and "dwelling," drawing the analogy that ethos provides for a human community the safety and security a stall does for an animal. But as we can see from the above examples of lions and fish, there may be a nuance of meaning locked out of that image of a safe-keeping barn, with its explicit construction and presumed security. Ethos has a fragrance of wildness at the same time that it conveys accustomed and proper place. It is much more akin to our word "habitat."

One might infer from Lehmann's statement that the use of the word for animal dwellings preceded in time its use for human dwellings, customs, and character, and that these uses represent a substantial change in the meaning of the word. This is not the case, however. Hesiod, who is generally placed in the same period of Greek literature as Homer, used it for human homes, for customs, and for disposition or character, all in the same work.[5] This multivalence of meaning continues through time. We have mentioned Herodotus's use for the homes of lions. Writing centuries later than Homer and Hesiod, he also used it for human homes and customs.[6] And, finally, Aristotle, who clearly applied the word in a moral sense with regard to human beings, also continued to use it with reference to animals.[7]

The origin of the English word "ethos"—and the word "ethics"—thus intimately links place, otherkind, and morality. "Ethos" in common English retains a hint of this confluence of meaning, in that its collective sense—the character or characteristic spirit of a social group or movement—is often applied to a community in a particular place. The connection with animals is also not entirely lost. The study of human ethos and its formation is called "ethology," a word whose definition includes the scientific study of animal behavior by observation in its accustomed habitat, as opposed to laboratory experimentation.

But is this just a peculiarity of the Greek language (hanging over in English), or does it represent some insight into moral life? The multivalence of ethos in Greek, particularly its connection of ethics with place, recognizes that moral subjects are by definition subjects *in a world*, not entities in isolation, existing in some immaterial void. Even if such a detached existence were possible, questions of morality would have little point, since morality deals with patterns of relationship. An ethos is that which provides, as William P. Brown notes, "the position and orientation of the moral subject vis-à-vis the environment," the natural-social context.[8] Ethos orients the subject to his/her location in terms of meaning, obligation, and value.

The concept of ethos as a network of norms also recognizes that a particular moral norm or ethic does not stand alone, but—as in a physical habitat—it exists as part of a web that is more or less coherent (and when less coherent can result in more moral conflict and confusion within a group). Any ethic proposed must deal with the norms already in place in a given ethos, and demonstrate either its adequacy in terms of "fit" with them or the inadequacy of the present constellation in some way. Not only does ethos thus provide "the sustaining environment or context for an ethic to function," but it also furnishes the context "for a moral subject to perform. The normative claim of a particular ethic and the integrity of the moral self are determined in part by the place they assume in the larger ethos."[9] It is this context in which we know who we are and how we should act.

"One is a self only among other selves. A self can never be described without reference to those who surround it," according to Charles Taylor. And this includes our selves as spiritual and moral beings: "We first learn our languages of moral and spiritual discernment by being brought into an ongoing conversation by those who bring us up."[10] Moral formation and discernment, then, are performances analogous to speaking a language, and so Ferdinand de Sassure's analysis of *langue* and *parole* with regard to the practice of language is helpful to a certain degree.[11] Sassure identifies the langue as the structure and fund of a language, common to all speakers, which is drawn upon and recreated by each particular speech act (parole).[12] An ethos, like a language, provides this structure and fund for the performance of moral life by a subject. The problem with Taylor and most other communitarians is the assumption that the "other selves" in the community in which a self is formed are only those of humans, that humankind alone generates a langue. Most also assume culture to be entirely self-generating.

The metaphor of moral habitat expresses the sense of a sustaining environment more holistically. It includes more than human members in the community that forms and is formed by an ethos. This entertains the possibility of moral import within and a morally formative role for what we generally mean by the word "habitat." In the case of Taylor, he is forming his argument against both the more reductive forms of naturalism that would categorize moral behavior as instinctive, and the highly individualistic assumptions of modernity. There is simply no need, however, to frame this entire question in terms of either-or, either moral nature as instinctive ("natural") or as strictly cultural. The dichotomy disappears when humankind is understood to be part of the natural world and that what is deemed cultural is itself coproduced by humans and the rest of the biotic community.

Ethos Formation in a More-Than-Human Community

That the morally relevant community is more-than-human is hardly a new idea; it has been gaining momentum in Western environmental circles since Aldo Leopold proposed a "land ethic" that "simply enlarges the community to include soils, waters, plants, and animals, or collectively: the land."[13] (That it takes an "extension" to our ethics to include this community says something in itself, however, about our ethics.) Many of the efforts to pursue this extension or land ethic still do not reach the point at which it "changes the role of *Homo sapiens* from conqueror of the land-community to plain member and citizen of it."[14]

One reason for this, I believe, is that the present state of philosophy and of environmental or ecological ethics spends most of its effort on one side of this forming/formed-by dynamic: what human social constructs make of the rest of the natural world. In ecological ethics the concern behind this preoccupation is the material impact of human activities on the ecosystems around us and the support of these activities granted by an ethos. This concern is understandable and I share it, since the activities of at least some humans now have reached such magnitude that they are altering major life systems of the planet—climate patterns, water systems (e.g., aquifer depletion, ocean life), and life-maintaining topsoil—at an unprecedented speed and scale compared to that of most of human existence.

And at the same time as these concerns have come onto the ethical agenda, it has become clear that "people do not live in raw nature so much as in their pictures of nature, nature as humanly imaged and 'cognized.'"[15] The perception of the rest of the land community is always shaped by cultural perspective. So that "nature" as an objective reality cannot be experienced, even if it exists, apart from cultural projections upon it. I am not denying that "nature" is always "cultured" for human beings, but I am insisting that the reverse is also true, that "culture" is also always "natured" to some degree.

Without addressing the questions of what the rest of a biotic community makes of humans (in this case, human ethos) in addition to what humans make of it, the emphasis remains on the humans as actors and on human meaning making, as if this capacity erupted in discontinuity from the rest of the biosphere. Of course we need to attend to what we are doing. But this preoccupation with our own impact needs to be balanced with attention to how we are acted upon if we are to realize that "the world, even as nature, is not an external, monolithic object to be handled, whether reverently or abusively, by detached subjects; it is at its core a community."[16] What kind of "community" would be constituted overwhelmingly by passive, voiceless members, who are primarily objects acted upon and not actors themselves? And what does it do to human moral formation to consider humans the only subjects in a world of objects, subjects marked—in a way that excludes all others—by moral agency?

While the metaphor of moral habitat has many features to explore, this is the one to which I will give most of my attention—the way it integrates the agency of the rest of natural world in our understanding of the moral life of a community. Like William P. Brown, I am less interested in "examining the material impact of human beings on the physical environment," than I am "in the reverse relationship, in the environment's impact, as conveyed by certain codified traditions," on a group's identity and moral character.[17] Among the relatively few scholars who are now attending to this side of the forming/formed-by dynamic, Brown explores it as part of his larger work in biblical creation texts, and Daniel T. Spencer introduces it into his enlargement of the social ethical idea of "social location" to "ecological location." Ecological location "acknowledges and places at the heart of ethical analysis nature's active agency, both as the whole ecosystem and biotic community, as well as its constituent parts of individual creatures, species, and niches."[18] Spencer also suggests that "how we are shaped to see and act in the world," in other words, ethos and agency, "results from a complex interplay of physiological, social, cultural, *and environmental/ecological* factors."[19] In the context of highly anthropocentric habits of thought, we may need to put an amplifier on that interplay in order to "tune in" to the ways in which the environmental/ecological factors play into physiological, social, and cultural ones. The first step in this process is to examine how an ethos is constituted, the means by which it is produced and reproduced. From what and how does a network of norms and values arise?

A network of norms does not stand as something isolated from all other aspects of what is broadly considered "culture," but permeates the myriad cultural practices of a people: language; stories, from sacred myth to entertainment to gossip; arrangements of space (architecture, landscape) and time (calendar, festival); ritual and ceremony; contemplative practices; procurement and preparation of food and what foods are eaten or forbidden; conduct of bodies, clothing, gestures, and attitudes toward bodies; gender and sexuality; practices of healing and reconciliation; power relations; inheritance, ownership, and trade; practices of knowledge acquisition, verification and transmission; procreation and child-rearing; poetry, music, craft, dance, and imagery; games and other play; including the technologies employed at times in much of the above. This list is not the only possible one and is not meant as an absolute. It is meant to be as inclusive as possible of a wide range of cultural practices.

Some of the elements overlap one another; bodies, for instance, are engaged in many of the other activities (sexuality, dance, ritual) in quite specific ways, and language is used throughout. These practices intersect; they also both weave and reflect the previously woven patterns of an ethos. For example, by listing "the conduct of bodies" as well as ritual, I mean to draw attention to the ways in which bodies are situated and move in ritual. To have to bend down to enter into and practically crawl to move within a low-ceilinged space dug into the ground and lined with fresh-cut cedar boughs places a ritual participant bodily in a relationship to Earth and cosmos that is not the same as walking upright up steps and entering a vast cathedral nave. Not a word need be spoken. But the words spoken—"all my relations"—in the ritual of a sweat lodge heighten attention to what is already being experienced with the body, in the ritual but also in daily existence. Yet language also, of course, shapes perception of that existence.

The importance of language in relation to ethos is widely acknowledged philosophically and by "common sense" to some degree. This common awareness of importance of language in patterning values and norms is apparent in the emotion generated by language used for God, women, people of color, sexual practices and identities, and various ethnic/national groups and regional areas. The *structure* of a language is involved as well, something that becomes evident when comparing related languages to one another, but especially striking when comparing quite different language families. The heavily noun-based structure of English and its related Indo-European languages shapes a perception of the world as being made up of discrete, fairly stable entities. A verb-oriented language, such as the Algonquian group, emphasizes processes over things. Where most European languages use gender as a grammatical category, Mi'kmaw and its related languages use animation. European languages are notable for their concern with time, expressed particularly in English (although it is not the most extreme example) with its multitude of tenses. Mi'kmaw, by contrast, is more concerned with location and relationship, incorporating these in ways that are more subtle, complex, and pervasive than Indo-European prepositions and possessives. Different constructions are used if the subject being spoken of is present or absent, and common prefixes indicate association or kinship with the speaker and listener. Language permeates and patterns our thought; it directs our attention, prioritizing what is important in the world—distinct objects divided (according to gender) and events in time, or dynamic processes in relationship in space.

Not only are the vocabulary and structure of a language important, but so is the dominance of either orality or literacy in the use of language. David Abram argues that literacy itself affects our perception and experience of the "more-than-human world."[20] While a homeland and its places can be "storied" in many cultures, only scholars with literate practices would ever think of topography in terms of a "text" that is "read," for example, with all the distance between self and land that that implies. But as we will see below, language itself also arises in the context of environmental/ecological/bodily experiences. Because of the volume and intensity language has received, my purpose in placing it alongside other practices in this list is to remind us that it is but one among other forms of moral discourse that are possible in any culture, and it does not alone shape culture or ethos.

Religion is present in this list in the form of practices (narrative, ritual, imagery, healing, dance) rather than as a separate category for two reasons. One is that I am interested in how religious *practices* as well as ideas shape an ethos, and the other is that many nature-cultures and most religious adherents do not conceive of religion as something distinct. For them, religion is not a matter of a set of beliefs but a way of life. This is not meant to diminish its importance, just the opposite. I am trying to employ a way of thinking about cultural/social life that would be widely applicable to the diversity of human nature-cultures, in terms that would be recognizable and acceptable cross-culturally. This list is not perfect in this regard, of course. It is doubtful whether one such list or definition of culture could satisfy this requirement. But given the purposes of this work, one of which is attempting to find better paths to cross-cultural ethics, every effort needs to be made to improve the adequacy and appropriateness of how we think about other nature-cultures and the phenomena we are attempting to talk about when we say "culture."

Clifford Geertz and others have noted the problems presented by "the multiplicity of [culture's] referents and the studied vagueness with which it has all too often been invoked."[21] He offered his definition of

culture as one without multiple referents or "unusual ambiguity," a definition that has become highly influential in anthropology, sociology, and
culture studies in general: culture is "an historically transmitted pattern
of meanings embodied in symbols, a system of inherited conceptions expressed in symbolic forms by means of which men communicate, perpet-
uate, and develop their knowledge about and attitudes toward life."[22]

Faced with this definition, let me speak up for studied vagueness
and multiple referents. My concern is not only that this definition and
Geertz's explanation may too closely reflect our own cultural preoccupations, but in doing so assume a unidirectional process in ethos formation
that negates the role of otherkind. Geertz quotes Susanne Langer that
"the concept of meaning, in all its varieties, is the dominant philosophical concept of our time," and "sign, symbol, denotation, signification,
communication . . . are our stock in trade."[23] Indeed. *Our* time, *our*
intellectual practices have until recently tended to abstract meaning, symbol, and signification from sensuality and tangibility. It does not necessarily reflect the self-understanding of other nature-cultures nor does it
give a wholistic portrayal of the dynamics of culture, particularly the formation and function of ethos.[24]

We can, with such an emphasis, too easily slip into conceiving culture as what is conceptual and symbolic, a pattern that does not taste,
smell, sound, or feel like anything. Those become accidental rather than
essential attributes, and culture thereby independent of any activity or
physical setting.[25] Geertz was reacting to the position of determinist be-
haviorism, a "laws-and-causes social physics."[26] In doing so he reduced
the importance of "behavior" or "social action" to the fact that through
it cultural "forms" "find articulation."[27] (His actual work demonstrates
at times a more nuanced approach than his own articulation of his theory
and method, however.) If forms "find articulation" then they must somehow preexist behavior and practices. In this view, culture would consist
of an "essence" that is formal and abstract (conceptual and symbolic),
which is then imposed upon or expressed through the materia of everyday life. (Geertz is just as contemptuous of his colleagues who work with
"material culture" as he is of determinist theories and practice.) The flow
is unidirectional, from concept/symbol to action/behavior/practice. This
is simply too partial; such a flow exists, but the whole dynamic is much
more complex. Material cultural practices shape, not merely contain or
express, symbol and concept—and norm and value. And they are in turn
subtly shaped by and adapted to, not only shapers of, the biotic community in which they are enacted, a point explored further below. To say this
is not to endorse a simplistic biological determinism; it is to recognize the
complexity of ethos generation.

There is a related difficulty in Geertz's depiction of ethos, in that he conceives of ethos as distinct from worldview and assigns a mediating role to religion. Religion (any religion)—which he differentiates from "the common-sensical"[28]—has as an essential element, according to Geertz, the "demonstration of a meaningful relation between the values a people hold and the general order of existence within which it finds itself." Religious symbols not only demonstrate and store meaning according to Geertz, they actively "synthesize world view and ethos."[29] And this is the part that I find problematic. "World view," he says, is a people's "picture of the way things in sheer actuality are, their concept of nature, of self, of society." It is the cognitive and existential aspects of a culture as distinguished from the moral, aesthetic, and evaluative—the ethos.[30] The problem with this formulation is that Kant was wrong; the conceptual/existential and evaluative are not neat, separate categories.[31] But more germane to my immediate concerns is that the values that a people hold appear to have come from nowhere, from some vague, spontaneous generation that has happened prior to their "expression" as behavior or their "relation," via symbol, to ontology. This is not said explicitly, but there is no investigation of from whence they come or how they themselves are formed. Geertz is preoccupied (as are social constructionists as a group) with one side of the story. He quite rightly takes exception to the picture that humankind was fully biologically formed with its present attributes before the development of culture, pointing out that certain aspects of humankind—the brain and nervous system, incest-taboo-based social structure, and the capacity to create and use symbols—developed nonserially in a "period of overlap between cultural and biological change" in the evolutionary development of hominids. But even here his concern is solely with the cultural impact on biological evolution, rather than the reciprocal impact.[32]

We will return to this question of human biological-cultural origins below; the point I want to make clear here is that the image of a people possessing a full-blown set of values and meanings, then imposing them upon their environment and structuring it through them, or having to call upon symbols to relate otherwise unrelated categories of fact and value is just as misleading as the image of the "contractual man" of Locke, who is fully formed and functional before entering into society. Values, meanings, and norms inform practices such as those named above, of course. These cultural/social practices also serve to hold norms in place and communicate them, powerfully. But not only that. Values, meanings, and norms emerge over time and are constantly being modified within a people's tangible, sensual, material existence, its life within a larger whole. It is the wholeness of the "craftwork" of culture that constitutes an ethos, establishing "various compatible ways of perceiving the world and acting in it in appropriate ways."[33] The importance of "thick description" of a culture lies precisely in the wholeness of it, in its sensitivity to meaning, but not in an identification of meaning as the "essence" of culture.

This explains one of the sources of alarm about globalization of practices named above. The construction of "monocultures of the mind," to use Vandana Shiva's phrase, is via just such tangible practices. Her work traces the practices of industrial agriculture and biotechnology to demonstrate this.[34] Culture is imported *as practice*, and ethos thus disrupted or even replaced. Of course it is true, as Akhil Gupta and James Ferguson have argued in relation to globalization, that people are social agents, "who never simply enact culture but reinterpret and reappropriate it in their own ways."[35] Gupta and Ferguson rightly object to the simplistic opposition of an autonomous local culture and a homogenizing globalization. What their argument ignores is that the alteration of the local biotic community in significant ways by economic and political powers centered elsewhere—such as in Shiva's examples of forestry and agriculture—limits severely the options for local cultural continuity, creativity, and its very survival. Any concept of culture abstracted from nature misses this key point.

Despite Geertz's formal approach to culture and his distinction of the evaluative and the conceptual, his characterization of ethos adds an important element to Stackhouse's "subtle web of values, meanings, purposes, expectations, obligations, and legitimations that constitutes the operating norms of a culture in relationship to a social entity."[36] Unlike Kant, Geertz does not hive off the aesthetic and affective from the moral in his definition of ethos as "the tone, character, and quality of life, its moral and aesthetic style and mood."[37] Ethos is as much the atmosphere of a place and people's collective life as it is norms that are subject to being articulated. Ethos is internally related with, constantly shaping and being shaped by, *eros, mythos,* and *pathos,* as well *logos.* It is found in meaningful embodiment in a community.

I am asserting an "is," that an ethos *is* shaped by the larger biotic community, that these factors *are* genuinely at play, even in a latemodern/hypermodern ethos distorted by a fictional Nature/Culture dichotomy. But I am also insisting that there are a couple of "oughts" bound up in this assertion of an "is": we ought to be attentive to how we are shaped by the rest of the natural world, and those of us in a latemodern/hypermodern moral habitat ought to allow ourselves to be even more shaped by it. These two "oughts" are related. When we become more attentive to these factors in our formation, we can respond more adequately and more fully. Our ignorance of this dimension of ethos itself limits our ability to respond, but does not eliminate the dynamic. This is not a situation of a healthy, sustainable community with a healthy, sustainable ethos. As Stephanie Lahar has observed, when we sever human experience from its organic context, we may stop being *aware* "of the shapings and natural containments that a particular environment places around human practices and social structures. *But of course environmental effects do not cease to exist.* Instead, society is shaped by a fractured relation to the ecosystem(s) it inhabits, losing both characteristic bioregional contours and a sensibility for natural limits."[38]

The next task is to look specifically at how environmental/ecological factors can be understood as intrinsic to human physiology, society, and culture. We begin with the physiological as it relates to human cognition, to the development of the human brain and the way in which embodied existence in a world structures thought at a deep level, including our ethical thought. This will help us to engage more fully the ways in which environmental/ecological factors shape social and cultural practices. I want to emphasize that I am not considering these relations reductively or deterministically. On the contrary. Reductive theories such as Richard Dawkins's concept of the "selfish gene" accord all agency to DNA molecules, as if humans as organisms and communities had none.[39] This is the opposite of what I am saying. Humans as organisms (and possibly other organisms as well) *do* have moral agency, *and* we have it because this is, to use Lawrence E. Johnson's phrase, "a morally deep world."[40] I am arguing that we as persons and communities have this agency because of the agency of the larger biotic communities in which we evolved and are nurtured, not in spite of the nature of these communities, or in contrast to them. True, the agency of otherkind is not identical with moral agency as it is conceived in large part by the Western philosophical and religious traditions, a point to which we will return in the context of a discussion of the deficiency in Western concepts of moral agency.

Of course this is a value-laden interpretation from the outset. So is Dawson's idea that genes are "selfish" or his metaphorical assertion that "we are survival machines—robot vehicles blindly programmed to preserve" those "selfish molecules."[41] So are E. O. Wilson's similar statements that "human behaviour . . . is the circuitous technique by which human genetic material has been and will be kept intact. Morality has no other demonstrable ultimate function."[42] These are not objective assertions about "reality" but opinions steeped in Western instrumentalism. There is simply no value-free standpoint to take concerning the location of agency and the relationships of the rest of the natural world, human physiology, social realities, and cultures. (Such is the interplay of the biological and cultural, which I am not disputing.) When reductivist biology pretends to ethics, the argument from common sense about ourselves as agents at an organism level fits better with the actual events.[43] For one thing, the existence of people who forgo having biological children as a matter of moral conviction, sometimes against a prevailing cultural norm, cannot be accounted for in a reductive scheme without convolutions, adding exception onto exception. Nor can acts of self-sacrifice (particularly life risk) for the benefit of unrelated strangers.

We pay attention to reductive, instrumental theories partly because of the prestige of their proponents and partly because we intuit the validity of the idea that the origins of the biosphere, human origins, bodies, and moral lives all have something to do with one another. Human nature-cultures generally narrate their biological and social origins in a way that implies meaning, value, and normative behavior. The narrative of evolution in no way reduces human persons to automatons and human morality to strictly a means of genetic propagation. There are entirely different ways to interpret it to begin to attend to that connection.

As in most cases, it matters how you tell the story.

How Earth Made Us Human

When we think of the distinguishing marks of the human as a species, cognition and speech usually are at or near the top of the list. We tend to attribute anything qualifying as human culture to these capacities (along with our dexterous hands). We focus as well on the singularity of these abilities, emphasizing their *dis*continuity, rather than their continuity, with the rest of the biotic community. We (Westerners) rather smugly assume their superiority—that they are the evidence that we are of a higher order or, in the "religions of the Book," that we are "made in the image of God." What if we thought of every attribute that enables us to be culture making creatures, including cognition and speech, not as an imposition upon or an aberration in, but as a product of and response to the activity of a living world? It is important we examine more closely the genesis of these capacities for human culture making because we generally locate the origin of ethics in these capacities and most adamantly divorce our understanding of ourselves from the rest of the natural world on their account.

Niles Eldredge has remarked that in his conversations with creationists about evolution, he has been told that a primary objection to the theory of evolution and its story of the origin of humankind is that having so much in common with other animals would take away any source of morality.[44] Strict creationists aside, most Westerners would have no objection to our *physiological* origin in and kinship with the rest of the natural

world. The contention between religion and science on the matter of the
biological process of creation is only between a small group of people on
either side who seem to have a deficiency of imagination. Midgley is no
doubt correct when she says that most Christians (and I would add other
faithful people) today "readily accept that . . . God, if he could create life
at all, could do it just as well through evolution as by instant fiat. Many
would add that this more complex and organic performance is the greater
miracle."[45] Even so the gap between ourselves and otherkind is adamantly
maintained when it comes to the origin of human moral capacities. Many
of those people of religious faith who are untroubled by the idea of creation via evolution tend to rely on a vague and mysterious direct imprint-
ing of the divine image, the "divine spark"—identified early on by
Christian theologians with reason—to account for them. But why should
it be in any sense less miraculous that moral capacity should be created as
part of this "complex and organic performance" than for the whole to be
created in this fashion? Why would such a capacity need to be generated
apart from the whole? Why should human moral capacity be an anomaly
in creation? And is it really a function of reason? What is reason?

Reason as the location of morality survived, and thrived, in the rise
of the influence of secularity in Western thought, taking the place of God
as the transcendent reality. In the full flood of modernity, it was also a
property exclusively human (and the exclusive property of only certain
kinds of humans as well). From the standpoint of either religious or secular faith, what is often shared is not only the gap between reasoning,
moral humans and all Others, but the judgment about that gap. "That
our moral capacities are 'what separates us from the animals' is widely
seen, not just as a fact, but also as a necessary claim about their value,"
according to Midgley. She means here the value of moral capacities—
"Any doubt cast on their uniqueness is easily felt as an aspersion on the
reality and importance of morality"[46]—but this is only a difficulty because of the way we understand and value the rest of the biosphere, and
where we locate morality and agency.

We will return to many of the points in the discussion below to
draw out the implications of other-than-human agency in the next three
chapters and pick up again the thread of what it means for how we understand human moral agency specifically following that. The point here
is to begin to trace the connections across the conceptual gap of Nature
and Culture, to show how the earth has made and makes us human, so
that we can think about culture, nature, and ourselves differently. This is
one version of the story.

As Paul Shepard tells it, the story starts in the grass. The development of nutritional quality and quantity of energy stored in the seeds of

grasses and related plants holds latent possibilities of mind. This grass-and-seed energy fuels the herds of large grazing animals, who make the grasses' storage of surplus energy in their seeds worthwhile. Grazing prevents the forests from taking over; it maintains habitat for grass. With the gift of surplus energy in seeds, the grasses thus recruit herbivores into the process of their own propagation. And with the herbivorous herds, packs of *social* predators become possible. Mind, says Shepard, is "the child of the hunt."[47] Whereas at first the encounters of hunter and hunted has been largely by chance, "the stupid predator's random search and the stupid prey's contingent vulnerability," this nutritional abundance makes the "pursuit of the risky brain possible."[48] Memory, recognition of signs, anticipation, mental mapping of terrain, calculation of relative distance and speed, and deception, along with communication for demonstration of skills to the young and cooperation among herd or pack members—develop in predator and prey in tandem, evoked by the presence of the other, serving both. "This Cenozoic mutuality of mammalian hunter and hunted is one of the few long-standing and conspicuous episodes of reciprocal mental evolution."[49] Shepard bases this depiction of the development of the brain and mind on measurements of fossil crania over a period of more than forty thousand years and notes similar development in the sea. Since he is telling a story of how the earth made humans, however, the action followed is that on land. But humans are latecomers.

Even the predecessors of humans enter this drama and its development of mind when it is already long underway, "like Americans arriving, decades late, on the world's soccer fields." They join as both predator and prey, probably more the latter at first, but it is through joining what Shepard calls "the Game" that protohumans become humans. Their primate inheritance—social relations, a well-developed vocal system, and chimpanzee-sized brains—compensates for the comparative aural and olfactory disabilities of this "wily band of frog and cicada munchers, would-be meat eaters who . . . would parlay cognition into new realms."[50] Shepard describes the process by which early vocalizations associated with animals, plants, and places are transformed into words that marked the world as wolves would mark with urine. Through these words the minds of bipeds could carry the world in much the same way as their freed hands could carry bits of the environment. Human imagination becomes "more densely populated by recollected, imagined, represented, and dreamed forms than by tangible presences. A leap in mind was occurring in which meanings could have echoes in other realms, perhaps based initially on analogies between themselves and the other species, as when they danced the fighting. Humans tracked *into a new world of double meaning, based on an amplified relationship to plants and animals.*[51]

This "leap in mind" was fueled by plants and animals, and not just in terms of calories. They were food for thought itself. The becoming-humans "acquired the universal attention of omnivory, the soul of which was the prospect of an infinite world of latent meanings." Plants signal, among other things, types of soils (and the presence or absence of water), and, with their seasonal cycles of "sprouting, leafing, blooming, fruiting, quiescence . . . chronicle the year, keepers not only of their own periodicities but those of animals who depend on them," and so time is registered, future and past. At the same time, "the animal world provided models for the very idea of thought."[52] Animals in their similarities to and differences from one another provide a living scheme for developing mental abilities.

This is the other side of the story that social constructionists skip over. The capacity to create and use symbols is something that develops, over time, in concert with the living world, which both presents the need to do so and is composed of intrinsically meaningful entities and patterns, such as cyclical time. Take classification. Every animal must categorize its world in various ways: food, predators, potential mates, pack/herd members, species members, and so on.[53] Animals, in particular, according to Shepard, provoke classification based on their similarities to and differences from one another in appearance and behavior, and then further stimulate the need to accommodate ambiguity, since some animals defy easy classification (e.g., bats, who behave both like mammals and like birds). This is not just a phenomenon of physical necessity or a once-for-all evolutionary development. Human children who have animals in their vicinity demonstrate the continued relationship of that encounter to verbal and cognitive development. Shepard concludes his discussion linking language development in two-year-olds to cognitive development based on classification with the observation that "the mosaic of animal kinds is the supreme concrete model upon which this skill is achieved, and, as an added benefit, being alive, they keep before us an organic figure of reality, a world of kindred beings as the basis of a purposeful, living cosmos. The identity, names, and behavior of animals give us some of the first satisfactions of the mind."[54]

It is important to the concept of ethos as moral habitat to recognize that these similarities and differences do not dictate a particular order; indeed, the presence of ambiguities actively undermines the imposition of a uniform universal ordering scheme. Biologists continue to argue about taxonomy. Nor does it imply that meanings are fixed; what is provoked is the development of the cognitive activity of ordering and the *finding* (as well as making) of meaning. I say finding in addition to making, because while meanings are not fixed, they are not necessarily arbitrary.

Strict social constructionists have gone overboard at times in their claims. Something like a tree or a mountain or a bird or a skeleton can have multiple meanings,[55] but it cannot mean just anything at all—a mountain does not lend itself to signifying daintiness, while a chickadee might. It is also important to the idea of ethos as moral habitat that it is a *living* world of *kindred* beings that gives us the idea of a meaningful cosmos.

In this "new world of double meaning," into which humans tracked, where animals (and plants, topography, and weather, I might add) populate memory, imagination, and dream, we find emergence of art, narrative, ritual, metaphor—composed of and with this "living world of kindred beings." In the signatures of the animals, their signs, lay the origin of abstraction and symbolizing in a drama we primates already "knew" was social at heart: gestures, expressions, innuendos. In less than three million years, all these categories of the self and society were shaped by the traits of animals observed, the dangerous, competitive, beautiful, tasty Others.[56]

From this brief look at the evolution of the physiological (body/brain) and the mind (classification, signification, metaphor), we see how many cognitive capacities that make human cultures possible—perception, memory, forethought, imagination, communication—are not only shared by other species but have been shaped in a long, dramatic interplay with this "world of kindred beings."

And this is not the end of the story. I mean this both in the sense that humankind is not a telos of evolution and to indicate that human cognition, having once been developed, does not become a culture-generating process separate from the body or the biotic community in which cultures happen. Human cognition, even what the Enlightenment thought of as "Reason," has not transcended its embodied state in a biotic community. To grasp this goes beyond pinning together the conceptual rift between mind and body, while continuing to conceive of our minds working in much the same way as we did before. As Lakoff and Johnson put it, "the mind is not merely embodied, but embodied in such a way that our conceptual systems draw largely upon the commonalities of our bodies and of the environments we live in."[57] They argue that even abstract conceptualization and reason employ conceptual metaphors based on bodily domains such as kinesthetic experience and perception. Concepts we associate with reason work so well to "handle" our functioning in the world because "they have evolved from our sensorimotor systems, which have in turn evolved to allow us to function well in our physical environment."[58]

To connect this with the *moral* of moral habitat, Lakoff and Johnson point out that our very ideas of morality are grounded in experiences of bodily well-being, and when developed into "abstract moral concepts—

justice, rights, empathy, nurturance, strength, uprightness, and so forth—are defined by metaphors . . . We understand our experience via these conceptual metaphors, we reason according to their metaphorical logic, and we make judgments on the basis of metaphors."[59] Lakoff and Johnson demonstrate this point by showing how a metaphor such as Well-being Is Wealth generates a whole cluster of moral thought based on a kind of arithmetic: harm is taking away something of value or giving a negative value, which can only be redressed by the reverse, so that justice becomes a "settling of accounts" in which people get what they "deserve" or are "owed."

A more direct relation of bodily experience to moral concepts can be drawn with the related basic metaphors of Well-being as Health and Moral Power as Strength, in which an upright and balanced physical posture—the normal condition of a healthy person—represents rectitude. A virtuous person is "upright" and a morally approved way of life "balanced." Doing evil then becomes a "Fall." Resisting evil is "standing up to" it. Being unable to resist is to be morally weak. And an immoral person is "sick."[60]

Something like possession or wealth is, of course, culturally defined, although underlying any concept of wealth is the more general understanding of abundant sustenance for life. I am obviously not making a claim that culture is not a factor here. My tracing of the story of human cognitive development is intended to show that the capacity for metaphor itself, the need for and opportunity to abstract and symbolize, is generated in concert with the rest of the biotic community.

The weakness of Lakoff and Johnson's work is that it is heavily ethnocentric in the material they draw on for analysis, a weakness they observe. Their whole discussion of rights as possessions is certainly valid for the moral thought of the culture being analyzed and perhaps others, but it is questionable whether it would hold cross-culturally. That is not their purpose, however. They wish to show that even a culture that claims, especially in the wake of Kant, to practice an abstract form of moral reasoning, a tradition whose ethical discourse is oriented to principle, has at its core metaphor. Johnson makes the strong claim that metaphor is not only an inextricable element of our moral rationality but that it is "the chief means by which we are able to imagine possibilities for resolving moral conflicts, to criticize our values and institutions, and to transform ourselves and our situations." It is, indeed, "at the heart of our imaginative moral rationality."[61]

This characterization of moral reasoning as "imaginative" does, I think, hold great potential for cross-cultural ethical discourse, in spite of the limitations of Lakoff and Johnson's work. We will return to the moral imagination and its role in a moral habitat, but there is still an important

element of Lakoff and Johnson's work that should not escape us. In looking at what Lakoff and Johnson call the "source domains" of these
metaphors, we find a partial answer to the question of where the values
a people hold come from, in ways that would cross cultures.

Beginning at a very basic level, Earth has shaped the ways we understand well-being, and continues to do so at levels of more complexity
and diversity. I will explain this in relation to the most basic level first, although it seems patently obvious, because it must not be taken for
granted. Then I will explore the more creative and diverse ways the values a people hold are coproduced by habitat. At least some of the deep
metaphors identified by Lakoff and Johnson in moral thinking are based
on what people over history and in different places have commonly understood as well-being, at least for themselves: health being preferred to
sickness; purity to contamination of air, food, and water; sufficiency, even
abundance, to lack and want; social connection, caring, and nurturing to
isolation, neglect, or indifference.[62] If morality implies an understanding
of the good, these goods have set the most basic terms of that understanding and bind it together with the well-being of Earth, its health and
fecundity. These goods form the core of what Midgley calls "the most
basic repertoire of wants."[63] These are given by our existence as embodied beings of a social species; we share them with others of our own kind
and *also with many otherkind*.

Midgley helps us to begin to clarify the interaction of goods (in the
sense of what is deemed worthy, not in the strict sense of provisions),
wants, and cultures in the genesis of the values a people hold. Not all
wants are morally good, of course, and goods as well as wants conflict
with each other at times. Cultures take an active role not only in arbi-
trating these conflicts but also in subtly creating preferences for certain
goods and wants over others. Basic goods and wants are not the creations
of cultures ex nihilo, however. "We are not free to create or annihilate
wants, either by private invention or by culture. Inventions and cultures
group, reflect, guide, channel, and develop wants; they do not actually
produce them. . . . We cannot treat them as chance particulars, which
might be assigned any value and which we might decide to invent or discard."[64] What cultures do, according to Midgley, is coordinate, fix, and
develop systems of values rather than create values. "The notion of 'creating values' is a piece of nonsense—all anybody can do is adjust, develop, and extend them."[65]

The next question is whether cultures, beyond being a product of
human cognitive capacities and social nature developed through inter-
course with a world of kindred beings, and having being granted by Earth
a basic repertoire of wants that form the core of our notions of "good,"

then proceed to act wholly arbitrarily and independent of this living world as they group, reflect, guide, channel, and develop wants into systems of values. My answer is that it depends. This is where the diversity of human life begins to reflect the diversity of Earth. The degree of awareness about the participation of the rest of the living world varies from culture to culture, and the ways in which it does participate may be more evident and even more powerful one place rather than another. Cultural creativity also means that cultures differ in their response to the voices other-than-human members of a biotic community, each nature-culture comprising a distinct conversation, in its own language, or rather, langue.

Earth Conversations

Ancient and medieval writers of the West theorized effects of geographic location, climate, and topography in forming the character of a people, in ways that are at times quite sensible, but also include both the naïve and the highly objectionable. Physical characteristics such as skin color, physique, hardiness, longevity, and reproductive capacity, along with traits such as intelligence, belligerence, and righteousness were attributed to such things as altitude, waters, soil, and climate. Hippocrates' *Airs, Waters, Places* and Albertus Magnus's *De natura locorum* are prominent examples. "It would be difficult to overestimate," according to Clarence Glacken, "the amount of speculation about the influence of mountains, valleys, swamps, hard and soft environments" inspired by Hippocrates' essay with its sweeping generalizations about such influence.[66] For example, peoples who make their homes in well-watered lands are said to be "fleshy, ill-articulated, moist, lazy, and generally cowardly in character. Slackness and sleepiness can be observed in them, and as far as the arts are concerned they are thickwitted, and neither subtle nor sharp," while rough, waterless lands produce those who are "energetic, vigilant, stubborn and independent in character and in temper, wild rather than tame, of more than average sharpness and intelligence in the arts, and in war of more than average courage."[67] Or so says the Greek Hippocrates.

Given that such ancient schemes and habits of thought undergirded the European and neo-European constructions of "race"[68]—with all the evils of such constructions visited on bodies, minds, souls—it may seem foolhardy if not repugnant to broach the idea of environmental influence on a people, especially its moral networks. I am not going to make the kind of categorical statements that the ancients and medievalists, not to mention modernists, made with such seeming ease. But here, as with reductive gene theories, we have an intuitive sense that there is some link between climate, topography, biota, and human ways of life and character—or Hippocrates would never have found the audience he did, even if his ideas served powerful interests.

What Earth does is present certain opportunities for and restrictions on particular forms of human development. These are not uniform in all places. Local habitat shapes perception, form and ease of livelihood, population densities, and social structures. Societies are shaped by these opportunities and restrictions in many of the cultural practices listed above. It is not a matter of programmed cultural response to environmental stimulus, however. The conversing of the human imagination in the living world creates a human moral habitat. This imagination is not a capacity that, once formed, is disengaged from context. It is continually formed, stimulated and nurtured (or restrained) by the environs in which human societies dwell and which they craft with the energy, images, and bodies of the earth community. Earth is, in so many ways, continually forming as well as being formed by, human moral imagination.

Consider how this dynamic works in relation to the development of agriculture. Agriculture requires soils created by rock, weather, plants (lichens), and animals—as well as plants nutritionally worth cultivating and reliable sources of water. But these can take many forms, with varied consequences. Westerners tend to associate the development of agriculture with the "Fertile Crescent" and emphasize its role in the rise of settled populations, cities, and hierarchies—when surplus food made it possible to engage whole groups of people in non-food-producing endeavors. The fixation on our own origins ignores the independent (if somewhat later) development of quite different agricultural techniques practiced by forest peoples such as the Kayapo.[69] In rainforests, the productive layer of topsoil is thin, easily depleted, so they used a sophisticated system of swidden cultivation that required the reversion of planted areas to forest after just a few years. These swiddens were relatively small areas, cultivated on a rotational basis (sometimes a decades-long rotation), and spread over a large area, in concert with practices that built up the soil in between nonforest plantings. This apparently has been practiced sustainably for thousands of years. In contrast to the styles of agriculture made possible in the broad valleys of Egypt and Mesopotamia, this form of agriculture did not lend itself to permanent settlement or stratification of society. The point is that the difference in the topography and soil character shaped both livelihood and social structures.

Yi-Fu Tuan describes how a dense rainforest environment also forms capacities for perception itself and thus shapes cultural practices. A rainforest dweller is immersed in a relatively undifferentiated environment. Very little can be seen from a long distance; everything is seen at close range, and shades of green dominate. The forest canopy obscures

moon and stars, and seasonal fluctuations are minimal. So vision
becomes acute at close range, and the cues for perspective—relative size
of a visual image signifying distance and not just the size of an object—
are not (in Tuan's term) "read." It is sound instead that becomes the cue
to distance, location, speed, and size. And so one dances with an embrac-
ing forest of subtle rhythms and meaningful sound, rather than locating
oneself in an expansive cosmos. For the BaMbuti of the Congo, this
means that songs and music, not surprisingly, become an extremely
important element, even the focus, of rituals. And it is the sound of the
song, rather than words, which is important. Closeness with the forest,
embrace, is acted out in the practice of initiating a newborn, circling its
waist and wrists with vines to which are tied small pieces of wood. And
the location for lovemaking is in a forest clearing, rather than a hut.[70]

Tuan contrasts this inhabitation of a world with that of various
Pueblo peoples of the Southwest United States. Here, space, shape, direction, verticality, and color are the vocabulary of culture and value that
orient human members of the community. In a dry land, springs become
locations and foci of ritual, and rituals are patterned by a sun-marked
seasonality. Solstices determine planting and dancing, house building and
hunting times. Springs not only provide the source and continuity of life
in an area, they function spiritually and as a source of cultural identity.
The small spring near Paguate village recalls the original Emergence Place
and, in the description of Leslie Marmon Silko, it links "the people and
the spring water to all other people and to that moment when the Pueblo
people became aware of themselves as they are even now. The Emergence
was an emergence into a precise cultural identity."[71] The Emergence is
that event in which "all the human beings, animals and life which had
been created emerged from the four worlds below when the earth became
habitable,"[72] during which the human beings had to rely on the assistance and benevolence of antelope and badger. Silko explains that the stories of Emergence and especially Migration and their geographical
features create a "ritual landscape" for an "interior journey" of collective
self-consciousness, not an historical one in the modern sense.

The survival demands of the land required such a cultural-imagina-
tive journey: "Life on the high arid plateau became viable when the
human beings were able to imagine themselves as sisters and brothers to
the badger, antelope, clay, yucca and sun. Not until they could find a viable relationship to the terrain, the landscape they found themselves in,
could they *emerge*."[73] In the Hopi tradition, it is the very starkness and
difficulty of life in the high desert mesas they inhabit that keeps the people spiritually attuned. The clarity of the desert air, the vast visible dis-
tances, the extremes of temperature and scarcity of water magnify the

impact of each feature and creature. Nothing is overlooked or taken for granted. Each ant, each lizard, each lark is imbued with great value simply because the creature is there, simply because the creature is alive in a place where any life at all is precious." In order to survive in such a place "every possible resource is needed, every possible ally—even the most humble insect or reptile. You realize you will be speaking with all of them if you intend to last out the year."[74]

The importance of this shaping of perception and provision of the context/content of imagination by topography, biota, and climate, is that perception and imagination function together as a nexus of subject and world, funding the process of moral agency. According to Iris Murdoch, "I can only choose within the world I can see . . . if we consider what the work of attention is like, how continuously it goes on, and how imperceptibly it builds up structures of value round about us, we shall not be surprised that at crucial moments of choice most of the business of choosing is already over."[75]

Murdoch discusses moral perception and attention in terms of sight, perhaps to the detriment of the intimacy of sound, of conversation. Silko draws on both senses to situate pueblo people in a visually rich place of peril, in which survival itself is contingent on conversant relationship as well as what is "seen." Moral "imagination" should not be limited by its etymological root to the sense of sight, however, or we risk losing vital connections and dimensions of moral life. But if we expand the visual metaphor to include that of conversation, Murdoch's fundamental point is just as strong.

Moral imagination functions in several ways in the formation of a moral habitat, just as moral habitat (ethos) shapes moral imagination. One way in which it does so is the development of a capacity to empathize, to imagine what it is like to be Other, and to connect our emotions with such an imagination. A second function is the capacity to imagine ways in which conditions could be other than they are. How else could we live? Without some way of envisioning this, there is no sense talking about what is the matter with how we do live. Without a larger Earth community in which to reflect, we (humans or First Worlders) are captive in self-determined interests. And as Thomas McCullough points out, the moral imagination broadens and deepens the context of moral decision making in that it considers an issue in the light of the whole, by which he means not only the complex interrelated functional aspects of society—economic, political, and social institutions— but also "the less tangible but most meaningful feelings, aspirations, ideals, relationships."[76]

Pursuing the imaginative and metaphorical participation of the rest of the biosphere—and the indispensable role of the imaginative and metaphorical—in human moral life refutes the idea that there is anything "mere" about metaphor or imagination. They are crucial modes of knowing, operating even when we ostensibly discount them (or emotion) in favor of an ideal of abstract reason. As such, they are integral to the process of moral formation, personal and collective, and of moral reflection.

The point is not that other-than-human nature determines human culture but that it participates in it, that the cultural and the "natural" are so implicated together that even the capacities that we identify strongly with the cultural, including abstract thought regarding morality, can be traced across the conceptual divide we have put between ourselves and the rest of the biosphere. In the development of language, cognition, imagination, and social nurturance, in the origin of metaphor and its role in relating bodily experience into explicit systems of abstract moral thought—everywhere we find ourselves shaped by our embeddedness in the biotic community as a whole. Our spiritual, intellectual, moral capacities are "nature's own flowering in the form of us." They are "home grown" not alien.[77]

But surely this is as far as we can go in pulling threads together across the divide between human and otherkind. Other-than-human nature *in itself* has no moral content, apart from human beings, does it? The "morality" of evolution points straight to Social Darwinism, doesn't it? And even if Mind is the "child of the Hunt," doesn't the *moral* nature of mind consist precisely in that it "rises above" a Nature that is "red in tooth and claw," to reflect on, prioritize and arrange values? What about religious claims to revelation that transcends the created order? Fair questions. The kind of questions, however, that could lead us around the bend that makes a culture "in and of nature" run "full grain against it."[78] So they require care in how we seek answers. In order to do so, we will approach predation, evolution, and revelation from different directions in the following chapter.

Yet if we move on to consider the insights from cultures outside of the dominant Western worldview, leaving the discussion of the role of other-than-humankind in the realm of metaphor alone, we have created a barrier to taking their insights seriously. Moreover, the question of an "earth conversation," even if it would not be dismissed as completely imaginary (that is, unreal), is from that point on ultimately in the domain of the human. Human thought and language, and the langue of human culture, may be evoked in response to otherkind, but once taken out of the realm of actual encounter, there is little opportunity for ongoing dialogue. It is in moments of encounter that we come to know, as Adrienne Rich describes, that "the Great Blue Heron *is not a symbol . . . it is a bird. . . .* The tall, foot-poised creature had a life, a place of its own in the manifold, fragile system that is this coastline;

a place of its own in the universe. . . . *Neither of us—woman or bird—is a symbol, despite efforts to make us that*" (emphasis mine).[79] Verbal speech is, indeed, what "my kind of creature does" to acknowledge the being before us, but it is provoked by that being, and it is not the only form of speech or conversation taking place.

According to Donna Haraway, the world "neither speaks itself nor disappears in favour of a master decoder."[80] But what if it does speak itself—in a language in which we have lost fluency or mere competence in the process of developing our own highly specialized form of speech, or in our narcissistic preoccupation with the kinds of things our kind of creatures do, in our taking a stand conceptually "outside" the conversation, or even, as Abram suggests, in the process of developing written language? A return to the Sassurian concept of langue could help us think beyond speech as a human monopoly, beyond human languages of words and grammatical structures and open the conversational context to consider more-than-human participation as ongoing. Even human-to-human conversation is not limited to vocabulary and grammar. It consists of facial expression, posture, gesture, touch, and nonverbal sound. Both culturally specific meanings are conveyed, as well as meanings that can be shared across cultures. Given the power of such modes of communication, it is necessary to include them in the category of conversation. Whatever the capacities particular species (such as dolphins) may have for intraspecies communication, *inter*species communication continues, although humans are not always participants. Certain cultures lack fluency, or deny the possibility by defining communication in terms of vocal speech. But if we consider the langue of a biotic community to include gesture, expression, posture, and nonverbal sound, then every member of the biotic community can be understood as performing parole. We can think of each bioregion as having its own langue, dialect, or patois. And just as Taylor makes the analogy between language and moral discourse, we can begin to entertain a different approach to understanding moral formation and agency.

And finally, in moving to acknowledge the biotic community as formative, it is important to affirm that its relational nature is not merely communicative or imaginative, but it is emotive as well. Our sensual, embodied engagement with other-than-humankind can evoke strong passions to sustain our commitments to other beings and to the places in which we dwell and dream together.

Chapter Two

"The Great Community of Persons"

Northeast North America

Northeast North America was not the "forest primeval" or "wilderness" that Europeans thought they had "discovered." It was definitely not wilderness to its human inhabitants; it was home. I once mentioned to Isabelle Knockwood how impressed I was by the ability of the Mi'kmaq historically to navigate the forest. After all, in an adult education course in "wilderness navigation" I had taken, the instructor (who had lent her compass to a student for the final test of finding our way out to the parking lot) had herself become lost in the space of less than a kilometer. Isabelle reminded me sharply that people do not get lost in their own homes.

This land was hardly "untouched" in the way later Romantics may have imagined. Ecosystems are constantly in flux, and the indigenous human members of these ecosystems, like the other members, were and are both causes of and respondents to short-term, long-term, and permanent change. Over the course of ten thousand years, the Northeast North American continent had changed dramatically, from postglacial tundra to boreal and Appalachian forests, and toward the west, open prairie. Species of large animals such as musk ox had come and gone. Whether this extinction was caused primarily by climate change or overpredation by humans is a matter of speculation. But there is no question that human activity—including, in the more southerly reaches, initiating forest fires and practicing agriculture—was part of the ongoing life of the land. And changes in the life of the land meant changes in human activity.

In his ecological history of New England, William Cronon describes this as a dialectical process in which "environment may shape the interactions and choices available to a people at a given moment, but then the culture reshapes environment in responding to those choices. The reshaped environment presents a new set of possibilities for cultural reproduction, thus setting up a new cycle of mutual determination."[1] This cyclical description helpfully counters the idea that a culture develops away from or departs from a static "nature." It seems to be inadequate, however, to describe the dynamism of indigenous nature-cultures from anything approaching the ways it might be conceived from within their natural-cultural worlds. Cronon's way of describing natural-cultural dynamics posits an all-encompassing "environment" in a bilateral relation with "culture." Richard J. Preston presents a more indigenous conception in his study of the Cree with his phrase the "Great Community of Persons."[2] The world is not divided into a human society surrounded by an all-encompassing, unified "environment" with which human groups interact, but consists of many groups of beings, who are by and large all actors in multilateral relationships. In this section, I will discuss the natural-cultural history of the Mi'kmaq and draw on cultural practices of hunting, narrative, and language to analyze how certain aspects of ethos of this Great Community of Persons, particularly those dealing with other-than-human agency, are constituted within it.

Change has accelerated for the land and its people enormously with the occupation by European peoples. Of all the peoples of North America, the Mi'kmaq are among those who have endured the longest contact. Although there were early conversions to Christianity under French missionaries, and some attendant disregard for or active undermining of their religious understandings, intensive attempts at acculturation did not take place until the twentieth century. Until then the most devastating factors for the survival of the Mi'kmaq were the exclusion from their traditional lands by the English, destruction of habitat for other animals and plants, and most significantly, the "Great Dying," pandemics of eurogenic diseases. Yet through all of this they have survived as a distinct people.

Their situation is one of both rupture and continuity, and continuity even within rupture. In a colonialist situation, the threat of loss of cultural identity, practices, and values creates a heightened consciousness of them, a will toward preservation, even reification "as the need to protect [the] most central, tangible and significant elements grows."[3] According to Marie Battiste, although reserves may appear to outsiders to be acculturated to white society, they are not truly assimilated. "They are distinct cultural and linguistic entities who have survived the torture, rigors, and challenges of Christianity and civilization, while remaining loyal to their traditional customs, traditions, language, beliefs, values and attitudes."[4]

In her essay, Battiste describes how Mi'kmaq traditional cultural elements become reconfigured, but persist, in practices through which their Christianity is observed. She illustrates this with an explanation of how the gathering on the Feast of Saint Anne in Chapel Island (the location of a traditional gathering) continues aboriginal practices. Lisa Philips Valentine has discovered another means by which cultural elements survive despite great external pressures in her research on the "distinct Englishes" of First Nations people. While having been forced to adopt the language of a dominant culture, people nevertheless have "adapted these languages to reflect the distinctive histories and patterns of their culture."[5] And, of course, there are many elements—stories, practices of hospitality, healing, and other communal practices—that are part of the ongoing life of communities and transmitted from one generation to another.[6]

The question of which practices were aboriginal (that is, precontact or presettlement) as opposed to those that have become part of Mi'kmaq culture as a result of interaction with whites, and how these changes should be viewed, is a loaded one. When Mi'kmaq people debate this, it is often for a much different purpose than when white scholars or courts try to make a determination (which is not necessarily in Mi'kmaq interests). It is not my intention to make too fine a distinction between pre- and postcontact practices for the purposes of this study, but to accept Mi'kmaq judgment of what constitutes their practices. Where some of their traditions have become obscured or lost, those Mi'kmaq who identify themselves as "traditionals" have turned for their recovery to other linguistically related or geographically adjacent groups, even when the latter may have been at one time antagonists. But similar international cultural exchange precedes European contact. The people themselves are the ones to make the determination of the extent to which these can be considered authentic for them.

The Mi'kmaq are a single language-group within the large Algonquian linguistic family. The languages of indigenous peoples of Canada east of the Rocky Mountains belong to five families, of which Algonquian is the most extensive, with as many as twenty-six languages identified as belonging to the Algonquian family.[7] These include nations whose territories prior to European invasion covered an area from Newfoundland and Labrador south to the Carolinas, and from there west to the Prairies (except for an area near the Great Lakes). Linguistic and other evidence strongly suggests they share the same origins in place and in a language designated as Proto-Algonquian. With increasing distance and adaptation to some differentiation in ecosystems, ways of life and the language itself diversified from this original unity. Because of this diversity, there are distinctions to be made among the nations whose languages belong to this group, and even within them; nevertheless, there are still many commonalities among Algonquian peoples.

In Canada, until and long after European contact, these peoples subsisted primarily by hunting, fishing, and gathering in the subarctic forests and plains of Northeast North America, supplemented in some areas by agriculture (most of the land makes agriculture as a primary activity impractical). In addition to language origins, one thing the easternmost Algonquian peoples (Wabanaki, including Mi'kmaq) shared with the more westerly nations living in boreal forests (e.g., Cree, Anishnabe) is a situation in which food sources are diverse and scattered, rather than concentrated. This shaped aboriginal cultural practices toward mobile, rather than sedentary, patterns of settlement. Survival depended on intimate knowledge of a great number of species and geography, as well as a wide range of technique in food procurement. Such need for mobility actively works against the formation of an ethos in which accumulation of material goods is a status-seeking activity.[8] In addition to the territorial range over which it needed to take place, there is also the factor of the constancy of food procurement. There is evidence that what storage of food took place was in the form of emergency provisions, rather than providing stores for an extended period of time (over winter), because, while some parts of the year were leaner than others, food procurement was a year-round activity. Having no single primary resource to harvest contrasts with reliance on a seasonal harvest of migratory food, or some forms of agriculture. The absence of the practice of food stockpiling or intensification of reliance on a single source may act against the creation of a leisure class or social specializations that create highly stratified societies.[9]

Habitat may thus be a factor in social structures and consumption patterns, and the values that accompany them. But I don't want to stress this too strongly, so that it appears I am making a determinist argument. I am not. I want to make the point that a sharp distinction of natural and cultural influences in ethos, or the more extreme view of culture as the only influence, is flawed. The coconstruction is so subtle and interdependent that it is nearly beside the point to try to determine where to draw a line between nature and culture. In the case of the Mi'kmaq, the need for mobility and the diversity of food sources are integrated in an ethos strongly based around the practice of gift giving as both social lubricant and basic economy. An explanation for this cultural trope of gift giving that would limit it to a form of reciprocal altruism—that these social practices of gift giving served as a kind of informal social insurance policy—is far too narrow. Earth and its inhabitants, members of the "Great Community of Persons" are all conceived as givers and receivers of gifts. This is a cultural construction distinct from that of Europeans', to be sure, but simply noting the difference doesn't begin to consider why things are different, or how they came to be different. The abundance of life in Northeast North America reported by the earliest European colonizers may have been exaggerated for purposes of their own interests, but was likely based to some extent on reality. There were also, however, lean seasons, and the bounty of life was not always at hand, so that it was never taken for granted. Given that this biotic community had included humans for thousands of years—humans who did, at times, significantly impact biotic communities—it may be that a culture of gift corresponded to, responded to, and sustained that flourishing. Fortunately, when the Mi'kmaq situation became one of near constant scarcity in the wake of European settlement, that culture of gift continued to sustain them as a people.

The primacy of hunting for aboriginal and postcontact Mi'kmaq and other Northern Algonquians makes their hunting practices significant indicators of the nature of their moral imagination. One characteristic notion of Algonquian peoples, and one widely shared with indigenous peoples throughout North America,[10] is that personhood belongs to a wide variety of beings, seen and unseen, and these beings are considered to have societies that parallel and interact with human society in a complex web of mutual responsibility and obligation. A person, whether human or other-than-human, is regarded as a self constituted in complex webs of relationship and linked to everything else through the sharing of *manitu* (also spelled *mntu*), sparks of life from an original fire. At the same time, each person is capable of directing his or her own behavior. Each is due respect, so that others are not entitled to infringe upon a person's sense of her/himself, or to manipulate or coerce a person. "Since all things have a common origin in the sparks of life, every life-form and every object has to be respected."[11] So when the need for sustenance involves the giving of life for life, it involves petition, apology, gratitude, and respect for self-sacrifice.

A hunter's success is not a matter of skill alone, but also depends upon his[12] fidelity to responsibilities and in some cases, upon a special relationship with a particular species. The prey tests the hunter's acumen and virtue by its elusiveness, but ultimately the animal is given to the hunter (by giving itself or with its consent through the agency of a guardian spirit or a wind) or it is not. The prey is not thought of as having been overcome in a contest of wills, or by the superior cunning of one species (human) over others. In return for the gift, the hunter incurs obligations, such as treating the body with respect, using all of what is useful, and handling the rest in a fashion which makes it possible for the animal's spirit to return to a new body. Observations of seasonal change and regeneration of life became incorporated into a belief matrix that understands this sacrifice as part of one's being, not an end to its existence. What apparently is dead in winter returns to life in spring, not only that, but a "tree does not die; it grows up again where it falls. When a plant or animal is killed, its life-force goes into the ground with its blood. Later it comes back and is reincarnated from the ground. The spark awaits renewal and its life-force never dies."[13]

Other conduct requirements of this hunting ethos are making a swift kill minimizing pain, and a prohibition against boasting of a kill as one's own achievement. According to Harvey A. Feit, there has long been an awareness among Algonquian peoples that human technology and skill makes possible the killing of many animals, and there is a restriction to take no more than what is necessary to live. These obligations and constrictions are not merely the condensation of utilitarian calculations, however. They reflect the Algonquian construction of personhood, which is shared beyond the human community. The primary point I want to stress here is that other animals are conceived as acting intelligently and having wills and idiosyncracies.[14]

These properties are recognized as belonging to many different beings, not just animals (humans and other animals) or spirits. Wind has been mentioned, and this characterization extends also to other elements or meteorological events such as thunder, geological formations, and plants. The world as structured by an Algonquian language is a world of actors, of persons. Most words about things Euro-North Americans might consider "inanimate," such as the land, landforms, and features are verbs. "This means that the land is always thought of as extending or varying, and that speakers focus on how they and others experience the place they live in."[15]

Animate and inanimate are linguistic categories in Algonquian languages. These do not necessarily correspond to European definitions of animation, and it is too facile to assume that animation has the same meaning across cultures. But, according to A. Irving Hallowell, the linguistic designation, when considered together with Algonquian beliefs, attitudes, and conduct indicates a cognitive outlook unified around the concept of personhood. Restricting personhood as exclusively human and defining "animate" in eurocentric terms obscures a significant aspect of this worldview. "[I]n the Ojibwa universe there are many kinds of reified person-objects which are other than human but have the same ontological status," a situation which Hallowell grants is "a radical departure from the framework of our thinking."[16] Intrigued by this, Hallowell pursued it further. The fact that the category of grammatically animate includes stones prompted him to ask an old man, "Are *all* the stones we see about us here alive?" Hallowell relates that the man reflected for a while, and then replied, "No! But *some* are."[17] So the category of personhood is not necessarily fixed either by language category or classification by appearance.

As might be expected, given this possibility for personhood, human-plant relationships are at times conceived in terms similar to the hunt, those of gift giving and reciprocal obligations. Certain medicinal plants are said to reveal themselves only when the seeker has the proper disposition, and the combination of proper action and attitude on the part of the harvester and cooperation on the part of the plant are required for the potency of the medicine. "[A] person gathering roots, leaves or bark for medicinal purposes pleases the life-force of each plant by placing a small offering of tobacco at its base, believing that without the cooperation of the manitu the mere form of the plant cannot work its cures."[18] Stories of how certain plants became known as good for food or medicine often involve the notion of the plant revealing or giving itself to the people. The Maliseet (the closest related nation to the Mi'kmaq) narrative of Corn portrays a dying wife transforming herself into corn. In the beginning of the story, it is specified that her hair is yellow, an indication that she is not an ordinary human being but the plant (maize) in human form.[19] She tells her husband that she can remain with him if he follows her instructions. After cutting and burning the trees around their wigwam, he is to tie her hands together with strips of cedar bark, then, without looking back, he should drag her around the clearing seven times. He does as he is told, but when he has completed his task, nothing remains of his wife. Grieving, he abandons that place.[20] Upon his return in the following autumn, he finds the place full of stalks of corn with hair (silks) that remind him of his wife's. There is no doubt in the narrative that the corn is his wife regenerated, and through continued cultivation, present.

In both this agricultural narrative and the hunting ethos, there is a recognition that death is required for the sustenance of life, a death that may entail violence: the infliction of a mortal wound, the destruction of trees, fire, dismembering a body, and so forth. Nevertheless, this story demonstrates a common narrative thread of self-gift, even self-sacrifice, transforming death into life for another, a sacred transaction in the continuity of life, one that demands of the recipient gratitude and respect. It is important to emphasize that this violence is not that of taking life by force or against the will.

This theme is commonly present in Algonquian and other Native American stories of the "discovery" of useful plants. But an interesting phenomenon with this story is its striking contrast to the Anishnabe version of how corn came to be cultivated, in which Corn is vanquished in a battle rather than being shown as a willing self-gift.[21] The Mi'kmaq story also varies from the Maliseet, in that corn is the food of the dead, won in a gambling game by a living father who sought to retrieve his dead son, but "given gladly" by the guardian of the land of the dead when it is won.[22] Whatever the explanation for this difference among linguistically related peoples, I think it indicates that the portrayal of self-gift of plants is not merely a motif of convention, but may convey collective experience, reflecting ecological situations[23] and ethos.

It would also be a mistake to confine the gifts given by otherkind in the discussion of Algonquian worldviews to what humans make of their bodies—food, clothing, shelter, and medicine. As bearers of manitu/mntu, otherkind also grant gifts of knowledge and wisdom, guidance and aid. Unfortunately, this aspect of indigenous worldviews has been seized upon, extracted, distorted, and exploited by non-Native "spiritual" figures. It has also been dismissed as naïve or Romantic, or exploited for other commercial and political purposes. Abstracted from community identities, practices, and responsibilities, this aspect of indigenous cultures can become mere commodity, a form of cheap spirituality. Community practices often involve rigorous disciplines in order to make one available to more direct interspecies relationships and able to receive these gifts.[24] And while the wisdom or guidance gained is personal, it is not solely for the benefit of an individual. The stories are communal.

Mi'kmaq narratives commonly feature animals as leading characters, portray human-animal conversations and shape-shifting interspecies transformations.[25] These narrative animals do not remotely, however, resemble Disney caricatures or pets. They are accorded much more respect, even when treated with humor, and are at times regarded with awe. Nor are all such stories confined to the mythical time of folktale. In Sunset Rose Morris's "Spring Celebration,"[26] for example, the narrator—whose description of her own setting includes a television—asserts that a conversation between an owl and a woman was an event witnessed personally by her mother, although it was "a long time ago."

Both Mi'kmaq who identify themselves as "traditionals" and those who identify as Catholic share the sense of the reality of a spiritual world as fully present and as full of presences as the material.[27] "Each life-form begins with a potential for being, and the life soul is transformed at birth into an interdependent essence encoded in the form. As it develops, the life soul finds allies in other forces and manitu all around it."[28] A crucial point for this study is that nonhuman presences are revered and propitiated (*not worshipped*) not merely because they provide physical nourishment or medicine needed for survival, but because they are "felt to be inherently deserving of such regard."[29]

The Mi'kmaq and other Algonquin people live in a world of Persons capable of moral action and choice about that moral action, inherently deserving of regard. Relations between humans and other-than-human persons are highly structured and involve courtesy, caution, mutuality, reciprocity, deference, and diplomacy.[30] Among the gifts they exchange are not only their bodies, but wisdom and guidance.

There is certainly a different construction of predation from Longfellow's "red in tooth and claw" in the depiction of the necessary taking of life for the continuance of life in the mode of self-gift, rather than the taking by force against the will of a victim. Is this merely a rationalization, a convenient cultural construction to assuage the conscience in the process of death dealing? Or does it represent some insight into the nature of life? I write this in the milieu of a nature-culture whose dominant myth of animal and plant life is still one of Hobbesian and Cartesian mechanization, embodied in the extreme by the industrialization of agriculture and the management of "factory-farmed" animals, out of sight and mind of most of the population that consumes them. Both food animals and plants are referred to as "domesticated," etymologically implying they are part of households, but this is no longer the case. Most urbanites and suburbanites in the developed world never encounter the live animals or plants of whom their packaged foodstuffs are the remains.

Some reaction against this takes the form of the kind of animal-rights advocacy that presumes vegetarianism as the sole moral option, not only removing humans from the matrix of life-feeding-on-sentient-life by which a good deal of life on this planet is sustained, but also subtly deprecating hunting cultures as "primitive," archaic, or a lesser mode of moral existence.[31] Vegetarianism is not the only alternative to mechanized food procurement in a mechanical universe. Neither is a "return" to a hunting-gathering economy, even if it were practicable, necessary; but we might nevertheless be cautious about too quickly dismissing the testimonies of peoples who actually encounter and live intimately with other creatures as convenient rationalization. Part of the Western queasiness with predation comes because "images of predation as the power of the strongest confuse our monkey politics and its endless skirmishing for power with food chains in ecology, making the false analogy of nature to violence and war," according to Paul Shepard.[32] But it's an old queasiness, as we see in the next chapter.

Chapter Three

Agents of and
Respondents to God
Hebrew Bible

The modernist dichotomy of Nature and Culture, which this study contests, has been phenomenally influential in biblical scholarship and theology. Only with these as distinct and opposed categories could Gerhard von Rad have declared that (human) redemption was not only something separate from the doctrine of creation but the primary concern of the biblical Yahwistic faith. Creation, he says, "performs only an ancillary function. It provides a foundation for the message of redemption, in that it stimulates faith."[1] He goes on to further reduce the role of creation to "but a magnificent foil for the message of salvation."[2] So we have not only a dichotomy, but also a dualistic hierarchy, and an extreme one at that. As Theodore Hiebert observes, according to von Rad "nature is not only separated from human culture, but it is regarded as subservient to it."[3]

This is not the idiosyncratic position of a single notable scholar. It pervades much of the biblical interpretation and theological "givens" of an entire era. Hiebert identifies these two assumptions—the dichotomy of nature and culture (or as it is more commonly considered in biblical scholarship, creation and history) and their ordered hierarchy of importance—as providing "the basic framework for most of what has been written by biblical scholars about nature in biblical thought" over the course of the twentieth century.[4]

A similar, related dichotomy structures most of the scholarly research into the ways in which the ecological situatedness of biblical communities (although not conceived perhaps in these terms) may have informed their theological and moral imaginations. This is the dichotomy between the desert and the arable land. According to Hiebert, this geographic dichotomy entails three crucial (mis)conceptions in biblical scholarship. First, that the desert and the sown are two separate spheres, with two independent sets of people, one set living a nomadic, pastoral existence, the other cultivating the land as sedentary farmers. Second, of these two separate spheres, the origins of biblical culture and thought are associated with the desert and contrasted with surrounding cultures based on sedentary agriculture (Canaanite, Mesopotamian, Egyptian).[5] And third, desert nomadism and sedentary agriculture represent developmental stages in human civilizations, so that there is an evolution from herding to cultivation of the land. Israel, in this view, brings a desert-formed sensibility to the later stages of its "civilized" existence.[6]

Together, these two interrelated dichotomies—the hierarchical dichotomy between nature and history and the geographic/social dichotomy between desert and sown—"have provided the intellectual framework for the widely held view of Israel as a desert society that conceived out of its nomadic existence an historical consciousness new in the history of religions, whereby the natural world was devalued and stripped of its sacred significance."[7] Biblical thought was hailed in modernity as an innovation in the midst of ancient Near Eastern nature-oriented religions, its human-history orientation and "disenchantment of nature" (to use Max Weber's language) judged to be a distinguishing mark of the "progressiveness" of biblical thought, responsible for the religious impetus for social freedom in Western culture.[8]

Such a judgment may turn out to be more revealing of these scholars' understandings of biblical historicity and attitudes toward "nature" and freedom than it is of the framework of biblical writers. Toward the latter part of the twentieth century, a number of biblical scholars began to question whether the view of human history as the central biblical concern to which nature as a separate category is subordinate, is actually the "primary irreducible datum of biblical theology," as G. Ernest Wright confidently declared it.[9] Much of this questioning has come about in the wake of the assertions of Lynn White, Jr., who in his well-known essay "The Historical Roots of Our Ecological Crisis" connected the reading of certain biblical passages (primarily Genesis 1:28) to the Western domination and exploitation of Earth. Instead of heeding Thomas Berry's suggestion to put the Bible "on the shelf" for some twenty years and attend to the "book" of creation in order to have a more adequate reading of scripture,[10] biblical scholars and theologians have created a "virtual cottage industry"[11] discussing the issue ever since.

Reading from a late-modern sense of ecological crisis need not become merely a search for an anachronistic concern for "the environment" on the part of biblical authors. It would indeed be anachronistic to put their concerns in such terms—just as anachronistic as the modern scholarly imposition of the dichotomy of creation and history. But people living in the land would have had immediate and long-term concerns about well-being and sustenance and were no doubt very aware of the relationship of their own well-being to the flourishing of other life. *Where* they were as well as *who* they were shaped their theological and moral thought.

I am not attempting a summary of "the understanding of nature in biblical thought." In addition to questioning the appropriateness of the category of "nature" to biblical worldviews, I am aware of the inappropriateness of expecting the diversity of biblical streams to have a singularity of thought in this regard. As William P. Brown observes, "It would be hard to imagine, for example, a single worldview that could include the evolutionary drama of the Priestly creation account in Genesis 1:1–2:3(4a) and Qohelet's musings on the wearisome repetitions of the cosmos in Ecclesiastes 1:3–8."[12] What I am putting forward here is simply a discussion prompted by some of the very recent reinterpretations being brought forth by scholars who are beginning to question the foundational assumptions described above, to one degree or another.[13]

Within that growing territory, I am confining myself primarily to the issue of how biblical authors may have conceived the agency of members of the rest of creation, and secondarily to the reconsideration of the ecological locations of biblical communities and the ways they are incorporated in ethos within the texts. These are not the preoccupations of most of the ecologically motivated scholarship in general. Questions about God's care for creation and the human place in creation dominate that discussion, the former as it should. My questions are much narrower. I am also further focusing this discussion on the Priestly (P) Creation-Flood narrative of Genesis, although I refer to related texts. My choice here is predicated on the importance of these particular texts in their canonical position, the influence they have had on our moral imaginations when it comes to thinking about creation (ourselves included), and their prevalence in scholarly literature with regard to my questions. There are many other promising biblical texts to which one could bring questions about ecological location and other-than-human agency; I will refer to these but not investigate them further for this work.

The explicit question of *moral* agency of other-than-humankind in the Bible has rarely come up. It is, arguably, an inappropriate question, given that biblical authors would not be thinking in the categories of modern Western philosophy about this, any more than they would employ a dichotomy of creation and history. Martin LaBar[14] does address this question, and although at first he sets forth a fairly broad definition of a moral agent as "a being who relates morally to others," citing Richard A. Watson,[15] he quickly reverts to a more conventional idea of moral agency in his evaluation of texts. LaBar confesses his own surprise that what in his opinion is "rightly characterized as a homocentric book" nevertheless contains "some indications that people of Biblical times imputed moral agency to nonhuman animals."[16] In his brief (four paragraph) discussion of the question, he merely lists the episode of Balaam's ass (Num. 22:22–33), the curse on the serpent (Gen. 3:1–15), and other incidents of judgment upon animals (Exod. 21:28–32, Gen. 6:6–7 and 9:5–6, and Lev. 20:15–16) and a plant (Mark 11:12–14, 20–25). Then he concludes, "Probably because of my own world view, I believe that the world view of at least some of the writers of the Bible included some concept of the moral considerability, but probably not moral agency, of nonhuman entities. I come to the latter conclusion because I find it difficult to believe that the Bible teaches anything about the moral agency of figs."[17] Apparently, he reasons that since the fig tree episode is obviously symbolic, any other texts that suggest moral agency, even those in another testament greatly removed in time, must likewise be strictly symbolic. Symbolism is "the key to all the passages" in the two paragraphs of examples he cites, he decides. "Non-moral agents were involved in the acts of God and man as symbols."[18] His conclusions further imply that symbolic meaning precludes any other. At least he has the candor to admit that his conclusion is heavily conditioned by his own worldview.

I don't mean to imply that the Bible should be read in a literal fashion, or that we can ever know with complete certainty what biblical authors had in mind, but if we hope even to remotely approach biblical understandings, it is precisely our own worldview—our constructions, dichotomies, and expectations—that needs to be, as far as possible, suspended as we enter that other world with the empathy of imagination. Only in that space might we discover the character of biblical portrayals of other-than-human agency. A more in-depth look at texts is necessary as well.

To do that, let us begin at the beginning. I will consider primarily the Priestly (P) Creation-Flood narrative. But first, let me situate P briefly in its ecological location. It is not possible to place or to date with any precision the composition or redaction of the P narrative. It is generally agreed that in its final form, P is the latest of the four traditional sources of the Pentateuch, although assigned dates of its composition range from the years immediately before, during, or after Israel's exile. There may be sections (the Holiness code or genealogies) that are much older, but most of the primordial narratives of Genesis probably do not belong in that category. In any case, these were texts by urbanites, whether they were in Jerusalem or Babylon at the time. The existence of cities with whole classes of people specialized in ruling and priestcraft presupposes a location where land is productive enough agriculturally and pastorally to produce food surpluses, some of which (in times when the cult functioned) took the form of sacrifices for the maintenance of Priestly families.

Due to the wide variability of rain distribution and its unreliability, agricultural success was, however, precarious, and in ancient times created regular famine crises and more social stratification.[19] Only the severest agricultural crises would affect the most elite within the Priestly class (a likely group for the source of this tradition), but the life of farmers was quite different. A bad year might mean debt slavery for members of a farming family, even, perhaps especially, if others still managed to produce. In protesting them, the prophets attest to practices of "joining house to house and field to field" (Isa. 5:8) and exploitation of the poor. The preexisting national-origin story (J) involving (drought-precipitated) slavery and divine liberation, wilderness experience, and promise of land was a countervailing force. The Priestly writings (Lev. 25) contain provisions for restoration of alienated land, as well as restrictions on slavery and for manumission, but the prophetic writings and even Deuteronomy (15) have an urgency absent from the measured provisions of Leviticus, where the relief cycle was fifty years, as opposed to the deuteronomistic seven-year cycle.

The Priestly world ideal is order, and their vocation to preserve and restore it sacerdotally. If the mental capacities required by classification are provoked by animals, by their similarities to and differences from one another, as was discussed in the first chapter, then the highly structured Priestly tradition represents in an exemplary way the incorporation of that element at work in an ethos. When the political world order is shaken in either impending or actual exile, this tradition nevertheless finds order (though not stasis) in the cycles and patterns of a grandly conceived cosmos (certainly one engaged on a grander spatial scale than that of the Yahwist). Genesis 1:1–2:3 functions as a cosmological prologue to the whole Pentateuch; it is not only about origins in a primordial history, but sets themes that carry through the narrated history of this ordered world and its creatures.[20] At various points in this primordial history, creation itself is threatened. The first of these threats we will discuss, as well as its resolution in the first of five world-ordering covenants.[21]

In the darkness of the opening scene of Genesis, "three central components or characters of the subsequent story are identified: *Elohim*, the waters, and Earth."[22] In the delight over the discovery of parallels in other ancient Near Eastern cosmogonies, scholars have compared and contrasted them with Genesis 1 and thought they had found traces of ancient mythological motifs of battles among the Babylonian gods, of Marduk's victory over the watery Tiamat, and evidence for the Bible's demythologizing trajectory, so that "the sources of power in what we call nature exist only by the will of one sovereign Creator and are not independent spiritual entities."[23] Depending on what "independent spiritual entities" is taken to mean—they certainly are not of the same order as *Elohim* and they are not nonmaterial (if nonmaterial is what is meant by spiritual)—such a statement may gloss too quickly over what they actually are and do in the story. William P. Brown argues that we are mistaken to read the elements as inert, passive recipients of God's activity. The God of Genesis 1 is a sovereign Creator but also a collaborative one. "Divine activity introduces an intricate network of interrelationships in which both elements and life-forms assume crucial roles in the process and maintenance of creation."[24] The following discussion of Genesis 1 owes a great deal to Brown as well as others.[25]

In Genesis 1–2:4a, the waters and Earth respond to God. They fulfill God's command by action of their own, becoming collaborators if not cocreators. First, the waters converge to reveal the dry land (v. 9). The verb form here, a niphal imperfect, is often translated as a passive jussive, "let the waters be gathered,"[26] but niphal forms also express reflexive meaning, "let them gather themselves."[27] The choice in English might be reflecting a theological preference, but the meaning of the waters gathering themselves is perfectly defensible. "God does not split the waters with a staff, but summons them with a word, and they obey."[28] Having been revealed, Earth is then summoned to "sprout sproutlings" (*tadšē* . . . dešeʹ, root *dšʹ* v. 11). The command is sufficient because "[t]he power lies in the earth's fecundity, poised to fulfill the word and realize its own productive potential."[29] In verse 12, the earth responds to this command with its own subtle variation, usually obliterated in translation, when it brings forth (root *yṣ*ʹ) the abundance of "sproutlings" asked of it: seed-bearing plants and trees. In the Priestly style, which revels in repetition, this variation is intriguing, especially as God repeats Earth's variation when the earth is addressed again (v. 24). Brown suggests that this verse, "Let the earth bring forth . . . ," together with the following verse "And God made . . ." (v. 25), indicate that "the creation of these land animals is the result of combined activity of the earth and the deity."[30]

The Priestly style in this account is "intentionally cryptic" in ways that entice and suggest.[31] What is suggested here is a cosmic conversation.

God not only addresses but appears to listen to creation. Michael Welker, casting aside the abstract idea and theological categories of creation, discovered that in this text (and the J account of Genesis 2), "[t]he creating God is not only the acting God, but also the reacting God, the God who responds to that which has been created. The creating God is open to being confronted by the freedom, the originality . . . of that which has been created."[32]

As with the first instructions to the land, the water is instructed to generate the creatures that populate it in "rhetorically precise and suggestive" language that again links the process and the products (v. 20). The waters are to "cause to swarm swarming creatures."[33] "Far from a chaotic force, the waters are conceived as a positive and active participant in the process of divine creation."[34] The same also holds for the luminaries in verse 15, commanded to "be lights . . . to give light."[35] Sun and moon are fit to be rulers over day and night. "Light, earth, and water actively participate in this methodical development of creation. Through the divine word, God enlists them to contribute to the ongoing process of creation."[36] And their work is judged as "good," $ṭôv$, beneficial, beautiful, and *right*. In the Hebrew language and the Priestly ethos, the aesthetic and the ethical are unified. The beauty of Creation, its order and fitness have moral implications.

The real mark of distinction of this account from its ancient Near Eastern parallels is its peacefulness. Earth, seas, and heavenly bodies are active entities, not passive, inert matter, but they respond to and work with *Elohim*, not in opposition. There are no battles here, no *Chaoskampf*, no violence. By contrast with the Babylonian story, "Genesis 1 is extraordinarily peaceful in its representation of creation."[37] This nonresistance on the part of the elements raises a question in itself. Even if it is granted that the Priestly creation account portrays earth, waters, and luminaries as agents for the good, is that sufficient to qualify them as *moral* agents? Perhaps not. There seems to be no opportunity presented for the elements to refuse, at any rate. Before we leave the question at that, however, a few observations need to be made.

This noncombatant creation does represent a choice, at least on the part of the formulators of this account. It is a choice against the prevailing mythos of the ancient Near East, of which they had ample knowledge. It differs not only from other neighboring cultures but also from other biblical writers. "Particularly in the case of the waters, God could have striven, more typically, as a warrior, rebuking (e.g., Ps. 104:6–7, Isa. 50:2b), defeating (Job 26:12; Ezek. 32:2–6; Ps. 74:13–14; Isa. 27:1) or at the very least containing the forces of watery chaos (Jer. 5:22; Prov. 8:29; Ps. 33:7)."[38] It could very well be that the reason for the choice lies less with making a statement directly about the nature of the elements than it does to express ideals regarding the exercise of power and the character of rulership in the Priestly ethos. Brett detects an "anti-monarchic tone" to Genesis as a whole, and 1:28 as counteracting the notion of a royal prerogative by extending it to the whole of humankind.[39] But if we take our attention off the human for a moment, we notice that in Genesis 1 "an omnipotent creator does not make a powerless and slavishly dependent creation."[40] The sovereignty of God is not tyranny. In the true climax of this narrative, the Sabbath (2:2–3), God steps back from creation, "letting it go in its own orderly goodness and creative continuation. Such is the way of cooperation and coordination."[41] As Welker notes, this applies to more than the human. "The Priestly account describes the whole creation, not only and not firstly the human being, as itself active, separating, ruling, and importing rhythm, as itself producing and giving life."[42]

The very least that can be said, if antimonarchism is the thrust behind the irenic and collaborative portrayal of creation, is that the commission to humankind of rulership over Earth and other creatures (vv. 26–28) needs to be understood as advocating a rulership of the same order,[43] engaging Earth and seas discursively to enable their productivity. If God not only addresses but appears to listen to all of creation, then humans, who are made in God's image, are to do likewise. If it is rulership, it is one of restraint, stepping back when necessary to allow the others to proceed for their own good.[44] Even if inadvertently, this portrayal dignifies creation[45] with the opportunity to contribute, and it does allow for the possibility that in the Priestly tradition, the elements of creation were understood to possess a capacity to act for good in response to God. The question of whether nonhuman creatures might also possess the capacity to act otherwise is clearer when considering P's version of the Flood narrative.

"The earth was destroyed in the view of God and the earth was filled with violence. And God saw the earth and behold it was destroyed for all flesh had caused the destruction of his way upon the earth" (Gen. 6: 11–12). This translation is Anne Gardner's, and she chooses to translate *šht* as "destroy" where most translations choose "corrupt." She does this, she says—pointing out the shift in meaning of the word "corrupt" since the time of King James—in order to emphasize that physical destruction, not moral turpitude or metaphysical degradation, is meant.[46] But moral and physical integrity are inextricably linked in the Priestly schema.[47] So it is more probable that both physical and moral annihilation should be understood here.[48]

Gardner is undoubtedly correct, however, when she insists that "all flesh" is not to be understood as referring solely to humankind, but includes "sentient beings of land and air" as well.[49] In 1958, Alexander Hulst upset what was at that time by and large a scholarly consensus that P included animals in the sphere of guilt for the violence and destruction of the earth in these verses.[50] Many others then followed suit. One of those was Claus Westermann,[51] who cites Hulst's conclusion that *kol baśar* (all flesh) is used in prophetic literature only with humans,[52] so that we can have no certainty that animals share in the guilt in the Flood narrative. Westermann adds that the subject of *ḥmš* (violence) elsewhere is invariably human. He explains the "rather broad notion" of "all flesh" in 6:12–13 (which he nevertheless constricts to human beings) as being similar to the "equally broad notion of 'the earth' in vv. 11 and 12."[53] With regard to that notion, he says merely that "[t]he subject, 'the earth' is saying that the whole of people's area of operation was corrupted with them."[54] He gives no reason why the earth (*hā'āreṣ*) should mean something different here from what it means in Genesis 1, in which it refers to the land that appears when the waters are gathered. If the notion is truly broad, then it is likely a more inclusive one, linking what has been brought forth from the land to the land, rather than an exclusive one, referring only to human society to the exclusion of the rest of creation.

In the P Flood narrative, the phrase "all flesh" is used thirteen times.[55] Where there may be ambiguity in some instances (6:12–13), this is most definitely not the case overall: "I am going to bring a flood of waters on the earth, to destroy from under heaven all flesh *in which is the breath of life*" (6:17); and even more explicitly, "And all flesh died that crawled (*rmś*) upon the earth, both of fowl, and of cattle, and of beast, and of every swarming thing that swarms upon the earth, and every human being" (7:11); similarly "Bring out with you every living thing that is with you of all flesh—birds and animals and every creeping thing that creeps on the earth—so that they may abound on the earth, and be fruitful and multiply on the earth" (8:17).

Exactly what was the transgression of "all flesh"? The most obvious answer is that the violence in question is predation that violates an original directive given to "every beast of the earth and every bird of the air and everything which creeps upon the earth," in Genesis 1:30 to take "every green herb as food." Gardner argues for this. "In other words, no dispensation was given to eat meat and thus no permission granted to attack or kill other creatures. That non-violence of all creatures was viewed as the ideal can be seen in Isaiah 11.6–8 where normally predatory creatures such as the wolf, leopard, lion, bear and asp live in peace with defenceless beings such as a lamb, kid, and a child."[56] According to Brown, in the P version of the post-Flood covenant (9:1–17), predation "has become part of God's regulation of human life"; predation "has become a necessary evil, for which the Priestly author gives no explicit explanation."[57] This injects "a note of ambiguity into the conventional practice of meat eating, construing carnivorism as a sign of alienation from the peaceful integrity by which creation was originally established. Noah's blessing is a legal remedy for the intractable nature of violence."[58] It seems indeed inexplicable, if carnivorism or ordinary food predation is the antediluvian crime of "all flesh," to endorse the identical behavior (in humans) after the destruction of the flood. Where lies the remedy, merely in God's acquiescence?

I suggest that the violence that was destroying the earth—indeed threatened to destroy all creation—prior to the Flood in the Priestly imagination was not food-driven predation, but wanton killing. This condition would more aptly warrant the "cosmic conclusion, judgment by unbounded water." (Note that the waters now act as the agent of God's judgment.) Wanton killing would fit the description of humans and land animals as having "caused the destruction of his way upon the earth" (6:12), the implication here being that there were no boundaries at all, each having corrupted/destroyed his "way." Humans and animals were "bound up together in this hot zone of depravity, effecting complete moral anarchy, the outcome of *unbounded* predation."[59] It is precisely the boundaries set which are both the remedies and the clues to the nature of the antediluvian condition. Human predation of otherkind is endorsed but circumscribed (9:4). Murder of human beings by other human beings *or by animals* becomes a capital crime (9:5). And God self-limits (9:11), as if once again deciding that restraint is the mark of divine rulership.

It could be the point of the Priestly Flood narrative that their contemporary world in which human beings sacrifice and eat meat was not always the case, a point that would serve as an orientation on an ethical compass. "That peaceful coexistence among the animals and humans once reigned has enormous ramifications for the Priestly tradents' contemporary world. Although lodged within the antediluvian realm of history, creation's peaceable ethos still generates a compelling moral force."[60] At the same time, the Flood story and its concluding covenant could be another episode in a narrative in which God and creatures collaborate to create a world—in this case, God "acquiesces" to creaturely initiated carnivorism, finding it moral, *within limits*. These two possibilities are not mutually exclusive, and the tension in the text cannot be denied. That tension is all the more notable when it is set in the context of the narrative's origin in the Priestly class, the very class that was autho-

rized to conduct the ritual slaughter of animals. Could it be that powerful emotions were evoked thereby, which made the boundaries around
the causing of death to animals a strong enough concern for them to incorporate it in their primordial narrative?

Whether that is true or not, what we can observe from this discussion of one strand of the biblical tapestry is that it distinctly portrays
other-than-human creation with agency that has moral dimensions. God
calls forth the active collaboration of waters, earth, and seas in creation,
and declares the results "good," in not just an aesthetic but also a moral
sense. That goodness is threatened by creatures, not only human creatures
but "all flesh," a category clearly including land animals and birds. That is
why they, along with human beings, suffer the judgment of the Flood.
Their capacity for knowing and acting for good (or refusing) is confirmed
in the covenant, which God makes with them along with Noah (9:10).

I began at the beginning, Genesis 1, with this task of inquiring
whether biblical texts indicate a concept of moral agency in other-than-
human creation. The answer is yes, at least here. But I do not think that
this is an odd, isolated point of view in the Hebrew Bible overall. A similar investigation into the J narrative would be, I believe, just as fruitful.
Hiebert's study demonstrates the ecological location of J and the ways it
is incorporated into the texts,[61] demonstrating, as Carol A. Newsom puts
it, that "Genesis 2–3 was written by someone for whom creation was not
an abstraction."[62] Animals are conceived (by Yahweh) as close companions of the groundling (adam), fellow creatures formed of the same common ground, who are, from the standpoint of this story "in no sense
lesser or lower beings."[63] At least one animal masterminds the transgression that sets the rest of the story in motion—and bears a similar fate as
the humans. A tree confers "knowledge of good and evil." Behind the
folktale motifs, there is a clear sense of companionship and solidarity
with Earth. There is no heavy ontological or moral line dividing humans
from the rest of the creatures. Even the ground itself bears the consequences of transgression (Gen. 3:17b).

There are other texts as well—in Job, the Psalms, and the prophets.
In Micah, for instance, aspects of nonhuman creation are called upon to
witness, to make moral judgment, in the case between Yahweh and his
people: "Rise, plead your case before the mountains, and let the hills hear
your voice. Hear, you mountains, the controversy of Yahweh, and you enduring foundations of the earth" (6:1–2a). This investigation of the moral
agency of otherkind in the P Creation-Flood narrative is indeed only a
beginning of exploring the possibilities.

We have thus far travelled through a personified world of wise, self-
sacrificing prey and a world of respondents to a divine creator, who also

responds to them. These portrayals may, to the late-modern mind, seem to be merely remnants of earlier cosmologies, a first naïveté, a state of mind irretrievable to those who have crossed some invisible cognitive line and entered a world forever "disenchanted." A concept of the moral agency of otherkind might be accessible and acceptable in myth or through faith, perhaps, but the modern ethos prides itself on having an empirical, scientific worldview, viewing "objectively" a world dis-spirited and depersonalized, identified as "reality." But is it? That we shall explore in the next chapter.

Chapter Four

The Continuum
Post-Cartesian Science

In the nineteenth century, some of the physiologists who performed vivisection experiments routinely severed the vocal cords of an animal before beginning their work, so as not to hear its cries. Conceptually, and to some extent perceptually and imaginatively as well, the natural-cultural transformations that made up modernity have "cut the vocal cords of the world."[1] There is not room here to trace these transformations in their full complexity, and that goes far beyond the purpose of this chapter. I want to highlight one factor that is often left in the shadows when the focus is on ideas of property and liberty, or on economics and the pursuit of knowledge, all of which, along with painting, sculpture, and literature, underwent wrenching shifts from the fifteenth to the eighteenth centuries. This factor is what Thomas Berry has identified as "the central traumatic moment in Western history," the plague known as the Black Death.[2] Robert S. Gottfried concurs, perhaps to the point of overstatement. "The Black Death," he says, "should be ranked as the greatest biological-environmental event in history, and one of the major turning points of Western Civilization."[3]

The losses were massive. Within the space of a few years (1347–1351), Europe[4] lost anywhere from 25 percent to 45 percent of its total population, in some local areas as much as 50 percent. In 1351 estimates were made for the Pope that calculated the dead at more than twenty-three million out of a population of about seventy-five million.[5] Various theories as to the origin and vectors of the plague have been advanced, identifying both social (patterns of trade and Mongolian alliances) and ecological possibilities. There was probably no single cause, but the plague resulted, at least in part, from the violent disruption, primarily due to climate change, of the ecological balance of Eurasia in the late thirteenth and early fourteenth centuries.[6] And, of course, the spread of the contagion involved the coordinated movements of microbes, humans, insects, and rats.

"The shock was immense, and the mechanics and commonplaces of everyday life simply stopped, at least initially. When plague came, peasants no longer ploughed, merchants closed their shops, and some, if not all, churchmen stopped offering last rites."[7] The immediate impact shredded communal networks, greatly accentuating a sense of individual isolation. If there ever was a condition of "war of every man, against every man" and the consequent need for social contracts (the fantasy of Hobbes), it likely arose in isolated incidents in the chaos of the plague, rather than in some primal, impossibly presocial existence. The initial pandemic was followed in waves by others, never allowing for the repair of the fabric of established practices as they had been. The plagues finally subsided after 1665, more than three hundred years later, in a different world altogether.

Over this stretch of time, the plagues altered even the perception of time, greatly foreshortening temporal horizons in a situation where death could occur almost instantly and uncontrollably. They both created a shortage of labor, so that labor-saving devices were more necessary, and depressed the need for food, shifting the value and understanding of land and social standing. The relationships of preplague feudal culture moved dramatically toward the class structures of modern capitalism.[8] After the first plague, art took a turn toward the macabre and grotesque. It "no longer showed the harmony between man, reason, and nature—the forms of God placed in their proper, natural hierarchy."[9] The image of Death was transformed from an "airy skeleton" to "a horrible woman cloaked in black, with wild, snakelike hair, bulging eyes, clawed feet with talons, and a scythe to collect her victims, whom she feeds to snakes and toads."[10] The certainty of the sense of eternal structure in the physical, social, and spiritual world of the late Middle Ages collapsed. The established healing practices and their practitioners were especially invalidated, many of them literally wiped out. Despite the generational loss of learned elites, the urgency to find answers and remedies opened up whole areas of innovation and practices of knowledge procurement.

According to Berry, "two directions of development can be identified—one toward a religious redemption out of the tragic world, the other toward greater control of the physical world to escape its pain and to increase its utility to human society."[11] Although these two thrusts diverge into what may be opposite directions—away from the rest of the natural world with a desire to escape it or toward it with a will to overpower it—they share a sense of locating power *in* that world, power hostile to the needs and desires of humankind. Both are attempts, in their own way, to overcome that power, religion through transcendence and the secular community through "its new scientific knowledge and its industrial powers of exploiting the natural world."[12]

It would be a vast oversimplification of history to attribute the contours of modernity entirely to the plagues, and that is not what I mean to say. The socio-ecological series of events set in motion by the plagues did not wipe the slate of Western culture clean, forcing it to start anew. Roger Bacon's celebration of science as a means to put nature "on the rack," after all, came in the thirteenth century (he was also arrested and imprisoned for it). It was more like a forest fire, which clears the land and provides space for latent seeds to sprout, grow, and take over the area. The plagues weighted the choices of which elements of its heritage European culture seized upon and developed in the aftermath, and which were rejected or ignored. And some new elements were added. The point is that the ethos of modernity, which bifurcates the world into Nature and Culture, is itself shaped by climate change, and the movements of microbes, rats, and fleas. I also want to point out that the claim that otherkind contribute to the "moral habitat" of ethos does not mean that their contributions have to be considered inevitably benign to human beings. After all, the contributions of human beings are not always benign, either, yet we consider ourselves morally capable (if not always moral) beings. Besides, ethos is never far from pathos. And the story isn't over.

It was in this milieu of scientific and industrial activity of the seventeenth century—as much as in the solitariness of his mind—that René Descartes conceived his famous epistemelogical approach, arriving at the highly individualistic "cogito, ergo sum" and the extreme dualism of mind and matter. The proliferating machines in this industrializing world became the means by which Descartes understood matter. Bodies were, Descartes concluded, machines made by the hands of God. The human spirit was the mysterious "ghost in the machine,"[13] which he had decided certainly existed based on the activity of his thought. He determined, however, that animal bodies had no such "ghost." His reasoning was based on two tests that he claimed could determine the authentically human (the presence of spirit) from any other creature, no matter how similar in appearance or behavior: (1) the capacity for meaningful language (not simply sound generation) and (2) the ability to reason. "For it is highly deserving of remark, that there are no men so dull and stupid, not even idiots, as to be incapable of joining together different words, and thereby constructing a declaration by which to make their thoughts understood; and that on the other hand, there is no other animal, however perfect or happily circumstanced, which can do the like."[14] Language and reason are the indications of the presence of the soul in the machine, according to Descartes, and the qualification thereby of moral beings. Hence, the nineteenth-century vivisectionists did not regard the animals under their knives as living beings, but as machines. Actually, I find a perverse sense of hope in the fact that they found it necessary to physically silence these animals. At some level, the pain of an animal was communicated in ways that made the experimenters uncomfortable, at least. Perhaps Cartesian thought was not as completely triumphant in actuality as we may give it credit for being philosophically. The other-than-human retained, though faintly, a power that bears on human moral agency, even among those whose ethos had educated them to deny it.

To a very large extent, however, Descartes's tests remain the yardstick for both human and moral being, and they frame the context of many investigations into animal consciousness and language, in which the onus of proof is on those who would claim, contra Descartes (who simply asserted his rather tautological judgment), that animals might have capacities for thought and perhaps even language. That the Cartesian paradigm still exercises a great deal of potency was noted by Mary Midgley in her review of Marian Stamp Dawkins's careful (too cautious for Midgley) book, *Through Our Eyes Only: The Search for Animal Consciousness*.[15] Why, asks Midgley, should the burden of proof not be laid in the opposite direction? Why do we not presume animals to have consciousness unless demonstrated otherwise. Based on our interactions with animals, our similarities to them, and the relationship of our origins, the proposition that humans alone possess consciousness ought to be the surprising position that needs defending, not the other way around.[16] The "yawning metaphysical gulf between people who are pure subjects and animals that are mere objects was a fantasy," insists Midgley.[17] And she urges the more realistic, Darwinian approach of actually paying attention to animals, as Marian Dawkins has.

It is not my intent here, however, to argue based on scientific evidence that other-than-human beings possess the cognitive qualities by which we traditionally have defined human being and moral agency, although there is a wide range of evidence upon which to make that argument, particularly when it comes to primates.[18] My difficulty with that approach is that it accepts largely without question the terms of Descartes and Kant, both of whose "reasoning" when it comes to otherkind and the ontology of human difference is more the rationalization of preconceived notions of difference and comparative worth. My particular question of what a barely emerging post-Cartesian science might have to offer regarding moral agency was instead initially prompted by the report of one particular scientist who paid a great deal of attention to animals, Roger Fouts.[19]

An incident occurs in the midst of his book describing his research with chimpanzees who had been taught sign language, research that held to his ideal of "science conducted with compassion and respect for the research subject."[20] While these chimps displayed many of the capacities that we associate with human beings, one event made me begin to question our definitions and theories of moral agency, not only in relation to otherkind, but in relation to ourselves as well. The chimpanzees that Fouts was studying spent their days on a quarter-acre island in the middle of a pond, which allowed them some freedom of movement and yet contained them (for the most part—two of them had tricked Fouts one day and hijacked his boat). Chimpanzees swim poorly, sinking fast because they lack the body fat necessary to have enough buoyancy in the water to swim, and this makes them afraid of water. They also have complicated social interactions, and introducing a new individual to a group can be difficult. One unfortunate individual was harassed for weeks after her arrival, until she was taken under the care of Washoe, the eldest of the little band. Tragically, Washoe couldn't protect the new chimp completely, and the young chimp was found drowned. This prompted the director of the facility to install an electric fence around the perimeter of the island. Some time later, another new chimp, Penny, was introduced, and that very afternoon she panicked when left alone with the other chimps and vaulted over the fence into the water. As Fouts was preparing to try to rescue her (a dangerous prospect, given that the panicked chimp could easily have dragged him under), Washoe rushed past him and leapt over the fence herself to rescue Penny, precariously holding on to a fence post and reaching out to the drowning chimp. Afterwards, he reports, "While Penny was calming down, I had time to gather my wits and to let the enormity of what I'd just seen sink in. Washoe had risked her own life to save another chimpanzee—one she had known for only a few hours."[21]

Is this moral agency or not? Briefly (because I analyze this more specifically in the next chapter), in Western philosophical traditions a moral agent is one who has the capacity to make a moral judgment and act on that judgment, to know the good and do it, or evil and refrain from it. An actor is deemed to be worthy of praise or blame according to (varyingly) his or her intent, object/means, or end. These are presumed to be subject to deliberate choice. The act needs to be uncoerced, with the agent possessing sufficient knowledge of the circumstances. Charles Taylor further qualifies the moral judgment involved as "strong evaluation," a capacity to form second-order desires (the desire to desire something). A weak evaluator simply weighs options; she is reflective, but only in a minimal sense, weighing whether one option or the other will bring her the most pleasure, for instance. A strong evaluator develops a language of evaluation and can articulate the superiority of a choice.[22] Taylor is discussing an aspect of human agency, and focusing on what makes humans unique. The question for me is whether moral agency—of humans or others—should be so defined, both on the grounds of justice for animals and for the sake of an honest assessment of the human moral life. As Frans de Waal says, "Animals are no moral philosophers. But then, how many *people* are? We have a tendency to compare animal behavior with the most dizzying accomplishments of our race, and to be smugly satisfied when a thousand monkeys with a thousand typewriters do not come close to William Shakespeare."[23]

It might be fairly argued that Washoe does possess to some degree language and rudimentary reason. But was she using them here? Washoe was not likely to have been making a strong evaluation, or to have made one in anticipation of the situation, any more than most humans who put themselves in harm's way for the sake of a stranger in an emergency are necessarily at that moment so deliberating, or would ever have expected (certain professions aside) to have to act in such a fashion. Human beings, who are capable of using language to describe their interior states to other human beings, often report that they take action in such circumstances without thinking about it. They simply respond to someone in need. Washoe did the same.

Now consider this report from a group of scientists studying sperm whales off the coast of California. At five in the morning, the crew called them to see a strange sight. Nine sperm whales had gathered themselves in a rosette formation, with their heads together, and their tails pointing outward, like the spokes in a gigantic wheel. A large oily slick from seeping blubber surrounded them. The scientists then saw the reason for this formation: circling the group were four killer whales. Like musk oxen, the sperm whales had "circled the wagons," presenting the formidable weapon of their powerful tails to defend against attack. Thus far, it was a common scenario of herding animals and predators. But then, something remarkable happened.

The killer whales succeeded in isolating one of the sperm whales and injuring it severely. But instead of abandoning it to its fate and escaping to safety, as most herd animals would do, two of the sperm whales left the rosette and, one on either side, led their companion back among them. This happened not once, but several times, and in the process, more and more of the sperm whales were crippled. As a result, by the end of the carnage—and there was more torn whale flesh floating in the water than the killer whales could eat—every member of the herd had been injured, and every one could possibly have died (the research ship followed the killer whale group).[24]

We resort to quite a lot of verbal and mental gymnastics to try to talk about what this kind of act on the part of Washoe or the sperm whales actually is, if we don't call it moral agency. We call it premoral or protomoral agency, particularly if we are talking about "higher" animals. Or we classify it as "instinct." Some scientific or, more accurately, science-inspired theories that acknowledge the origin of morality in other-than-humankind (and some that would extrapolate this to human moral systems) argue from the basis of natural selection that even seemingly altruistic behavior is a sophisticated, gene-based survival system.[25] In this view, the pod of whales, for instance, is likely all related; sperm whales are thought to be matriarchal, and this group could have been related as mothers, daughters, and sisters to one another. Their solicited protection of the injured ones would then be understood as concern for the perpetuation of genes close to theirs. The problem with this explanation is that the survival of the whales as individuals and as a group, including the perpetuation of genes, is made less likely, rather than more likely, by this behavior. More whales were put at risk than would have been the case if the uninjured had abandoned their sister. Genetic models of instinctual "premoral" behavior don't fit Washoe, either. Like many cross-fostered chimps raised with humans, Washoe identified herself as human. She had difficulty at first when introduced into a group of other chimpanzees. Washoe in sign language called members of her own species "black bugs" when she first encountered them, whereas humans were "us." Even if by the time of this incident, Washoe had reconciled with her own identity as a chimp, Penny was a total stranger.

Another line of reasoning could be more generous—toward primates and other "higher" animals at least. As "next of kin," chimpanzees and other primates might share something close to, if not the same as, human moral capacities. Or, if we learn more about cetaceans, they too might just be able to squeak over the threshold we have set for moral agency. But then do we draw the line just as thick and hard? What about the loyal family dog who sacrifices its life to protect a child?[26] That crosses species altogether, and the dog would have nothing at all to gain. Do we redraw the line at social mammals and disqualify the self-sacrificial behavior of some insects on behalf of the hive or nest because that is (supposedly clearly) "instinctual"? How many times do we erase, redraw, erase, and draw dotted lines before we get to the point where we start questioning why we're drawing these lines and what these smudgy categories are all about, anyway? (This is a question we will consider in subsequent chapters.)

The investigations of science that have managed to get out from under Descartes's thumb do not just expand the circle of moral agents, but challenge the very definition of moral agency. I want to suggest a further possibility. The subject of evolution has been touched on in earlier chapters. And I have mentioned and objected to the narrow kind of utilitarian relationships between evolution and morality represented by Richard Dawkins, E. O. Wilson, and others. Paul Shepard's tale of "how Earth made us human" hinted at a different interpretation. The story of evolution has been told as unrelieved competition even with the members of one's own species. It has alternately been told as the cradle of "strong attachment" and the "community concern" that is at the heart of morality.[27]

There is, without question, suffering and death in the neo-Darwinian tale of evolution. There is also an overriding, awe-inspiring drive toward the flourishing of life. Neither "accident," nor teleology, nor "intelligent design" provide truly satisfying explanations[28] for the magnitude, diversity, grace, and wonder of which we are a part as cohabitants of this "slim wet planet."[29] There is a magnificent freedom in the emergence of the truly new, its variations on possibilities (more to the point than "accident"), and yet it moves distinctly toward diversity, complexity, and relationality.[30]

The science of ecology, with its flows of energy/matter, demonstrates how much we are intricately woven within a dynamic process. Our bodies/our selves are only apparently discrete. Each moment our breath takes in the exhalation of plants and other animals. We eat their life to live, as other animals eat each other and plants, and as plants eat light and shape soil into forests and flowers. If we leave Descartes's dualism behind, we cannot abstract ourselves from this reality any more than we can step out of our skins. Together, cosmology, ecology, and neo-Darwinian evolutionary theory present us with a continuum of life that "seems to challenge the traditional religious sense of sharp ontological discontinuity between humans and the rest of nature. Evolution blurs the lines separating what we used to think of as distinct levels of being, making it more difficult than before to distinguish human from animal, and living from nonliving."[31]

This freedom of life to self-create its myriad forms—forms that develop diverse expressions and degrees of affinity, attachment, generosity, self-gift, sacrifice, compassion, joy, justice, and beauty—suggests agency and at least a moral potency at many levels—gene, organism, species, ecosystem, biosphere.

Borders, Crossings, and Uncharted Territory

I have spent most of this initial attempt to develop the metaphor of ethos as moral habitat—and the exploration of other-than-human and human

moral agency in that framework—in terms of the project's amodernity, tra-
versing the crevasse of modernity's "constitutional" divide of Nature and
Culture, and resisting any idea of the superiority of the (ostensibly generic,
but actually raced/gendered) human. I have done this and done it as clearly
as I can because, in spite of the witty discourses of cultural studies and
post(hyper)modernism, the rhetoric and categories of modernity are still a
dominant force to be reckoned with in the hyper(not post)industrial constructions of ethos in North America.[32] However, this does not sufficiently
take into account the "techtonic" movements that have already been causing a shudder under our feet, even opened up deep cracks, in the (for some)
comfortable ordering of a world into neat categories of nature and culture,
human and nonhuman. I'm referring here to the development of technologies in which the distinction of the biological and mechanical is increasingly
unclear. The most challenging of these technologies to ideas of moral
agency include information and networking technologies, artificial intelligence, robotics, and what the "digerati" call "jelly beans," those chips that
are small, specialized, capable of communicating, and embedded potentially everywhere in the built environment of industrial life. They are not
just in computers and cars but carry your balance on your debit card, and
will soon carry biometric data and record your movements on your
"smart" ID card; somewhat later, perhaps, they will record information or
dispense medication in your body. These technoscience practices also include other more explicitly *bio*technologies: the mapping of the human
genome, the creation of transgenic creatures, and other forms of genetic engineering. These two streams are not separate, and they especially converge,
mingling the biological and the mechanical/electronic/informational in the
technology of altering matter at the molecular level, in the various forms
of nanotechnologies.

Our ideas of what is real, of perception and knowing increasingly
take shape within and are shaped by a technoscientific ethos. In this "moral
habitat" the constituted categories of categories of nature and culture im-
plode.[33] The consequences of this kind of dismantling of conceptual boundaries produces at times a radically different program and outcome than
environmentalists and deep ecologists might have imagined when they challenged the artificiality of the categories. Instead of subsuming culture into
nature, all is subsumed into artifice. It is conceivable now that Descartes's
neat cleavage between machine and human is completely disintegrating, not
just in the cyborg and transgenic figures that Donna Haraway invokes,[34]
but in the development of machines that will simulate indistinguishably
from humans the capacity for meaningful language and the ability to reason by which Descartes presumed to identify the human. If machines are
"smart"—can reason, invent, and use language—are they human? Are they

agents? Are they moral? Theoretically we could program all the parameters of practical reason into machines that would be "artificially intelligent," meaning they have the capacity to learn and reason that matches or even exceeds human capacity. Along with the other technologies mentioned above, even the classic definitions of life—self-generated mobility and re-production—can no longer be used to distinguish between carbon-based organisms and systems based on other minerals, such as silicon.

Within the rarified domain of technoscience laboratories, Haraway and Bruno Latour have broadened the category of agent much further than I have done. Latour describes the air pump Robert Boyle used to demonstrate the reality of a vacuum, for example, as an "actor," one of a species of "inert bodies" whose capacities to demonstrate, measure, and signify are endowed with meaning. These mechanistic "bodies" lack souls, but in lacking self-will and bias are regarded as more, not less, re-liable witnesses than ordinary mortals.[35] They intervene in and arbitrate reality (both acts of and influence on agency), according to Latour. Not only can they do this in a superior way to human beings, they were con-sidered particularly superior to women, as Haraway points out. Women were excluded from the laboratory after they objected to the demonstra-tion of a vacuum that relied on the suffocation of birds as all the air was pumped out of a chamber.[36] Both Latour and Haraway are not primarily concerned with the question of agency per se but employ it in demon-strating the implosion of categories while building a case for the insepa-rability of political and scientific discourses from each other, and in Haraway's work, the inseparability of these discourses from technologies, gender practices, and corporate and military powers.

It is possible to think of these kinds of nonhuman "actors" as agents if we twist the continuum of agency we have discovered in the world of post-Cartesian science, and bend it back upon itself to form a kind of Moebius strip. The elements, biosystems, and creatures that have made us, have made us crafting creatures, designers and builders of build-ings and machines, which turn and shape our bodies, thoughts, imag-inations, and range of choices. As Winston Churchill once said of archi-tecture, "We shape our buildings: thereafter they shape us."[37] Thinking about agency of artifacts in technology offers us new metaphors to think about what knowing might be, and how even "inanimate" other-than-hu-mans could operate as agents in a collective sense. Even when individu-als are not "intelligent" enough to make judgments for the good, as systems they might indeed be able to act intelligently. Kevin Kelly ex-plains how this works with networks of "jelly beans" in the technoscien-tific domain. He likens the power of "dumb nodes" of these or ordinary desktop computers networked into a "smart whole" to the architecture of the human brain.[38]

But how far should we take this ascription of agency to technological artifacts? Kelly and his digerati cohorts have begun to speak in terms of being compelled to do "what technology wants." They imagine the creation of machines intelligent enough to outsmart their human inventors, capable of movement and self-replication, fulfilling all our philosophical definitions of life. These machines could even be programmed to have or to "learn" capacities that would fulfill our most stringent qualifications for moral agency. In a technocast (an Internet presentation) sponsored by the *Dr. Dobb's* ezine, Doug Hofstadter and other prominent digerati mused about the possibility of "spiritual robots."[39] Hofstadter sees no more difficulty with the idea of changing the substrate of life from carbon to silicon than with the idea that sentient, animate matter evolved out of inanimate matter. He frames the "evolution" of silicon-based intelligence in the most rudimentary Darwinian terms: "Our brains got designed by evolution, which is just basically tooth and claw, fight for survival. Why couldn't that happen again within the medium of the silicon or other media whatever it might be? . . . In fact the whole thing is not exactly a question of whether, it's a question of when."[40]

Bill Joy, cofounder and chief scientist of Sun Microsystems, is almost the lone voice in this assembled cast, and among the digerati, to challenge the view that these developments are inevitable. In a provocative article in *Wired*,[41] he raised alarms about these technologies in terms of the consequences of accidental and deliberate wide-scale destruction, because, as he describes, once developed these technologies could easily be appropriated by terrorists. In his portion of the technocast, Joy estimated the possibility of a "potential extinction event" (of humankind) at 30 to 50 percent, if these technologies are developed along the present trajectory.[42]

Kevin Kelly responded to this concern by making an analogy to raising children. As with our biological children, we need to allow our "mind children" to develop independence from us, so that they might reach their full potential. To do otherwise would be cruelty according to Kelly. We should not let our worry about the "out of controlness" of our technologies worry us, "in fact: unless we can worry about technology, it's not revolutionary enough. That's why we're worried now: we're dealing with technologies that are very revolutionary. The way I like to think about technology is that we should be aiming to train it to be a good citizen *[applause]*."[43]

That small sign "*[applause]*" is enough to make me want to beat a retreat back into some clearly cleaved world of nature and culture, human and machine, and hold on for dear life to what is natural and human. But I think the question is How do we do ethics in *this* moral habitat? What passes for ethics among the technoscience elite is entirely self-referential, in terms and method. Donna Haraway points out that the discussion about ethics in a textbook on genetic technology "mimes scientific analysis; both are based on sound facts and hypothesis testing; both are technical practices."[44] According to Hofstadter, there are no ethics "until you reach complex societies. There might be some small amount of ethics or morality in primates" or other "higher" animals, he admits, "But effectively it's tooth and claw, dog eat dog and so forth. Up until you have human intelligence." His companions are even less generous. Hans Moravec dismisses even the possibility of ethics at all; what we call ethics is merely "self-interest but at a larger scale." Again, Bill Joy dissents.[45]

Donna Haraway exposes the ventriloquist act of the "modest witness" of the early scientific age, the purveyor of false objectivity. And she asks the critical question, *Cui bono?* Whose interests are being served? She also complains about the inadequacy of discourse about technoscience, that "[f]or all their inventiveness in making fabulous natural/cultural hybrids that circulate fluidly in vast networks, many actants in genome discourse seem 'to be suffering from an advanced case of hardening of the categories.'"[46] Haraway claims that we need to face up to the ways our practices recreate such things as the seed, chip/computer, gene, race, ecosystem, brain, database, and bomb as the "stem cells of the technoscientific body. Each of these curious objects is a recent construct or material-semiotic 'object of knowledge,' forged by heterogeneous practices in the furnaces of technoscience."[47]

But there is something still making me uneasy about her dismissal of "organicism" and, I think, something valid in my desire to cling to the natural and human (not as mutually exclusive categories, but as distinct from other identifiable entities). Haraway's technoworld of critical discourse is still inescapably self-referential. Animals, plants, and minerals are granted agency only as "forced allies," given their muted voice only as they are processed through the heterogeneous practices of technoscience, as "materialized semiotic fields."

Perhaps we have finally come to the place where it makes sense to set some boundaries; not that these boundaries cannot be and are not crossed, but to help us think about those crossings more explicitly and deeply. These are, of course, constructed boundaries, and we should be explicit about that and the reasons for their construction. There are two sets of distinctions I wish to make. The first is between that which is instrumental and that which is not. The second is a distinction internal to the instrumental category. It involves the discernment of what is sustaining, involving the noninstrumental as a point of reference and bringing to bear the understandings of life in a moral habitat.

I think we can make distinctions between what is fabricated and instrumental, and beings with meaning not confined to our purposes. The first category is populated by objects such as the air pump, bomb, and chip/computer. It can also include ideas such as the *concept* of an ecosystem, but not the ecosystem itself. The land, waters, air, and organisms that inhabit them (including ourselves), the flows of matter and energy, while influenced by human action, are not our invention. These belong to the other, noninstrumental, category. We can and do manipulate things in this category in order to craft instruments, but they are not primarily instruments or merely potential instruments, waiting for our manipulation.

One of the reasons to construct these categories is to try to get some clarity about agency and responsibility, and to think about boundaries to our sense of license in a world where our manipulative powers have reached the point that we are largely and literally out of touch—because of the ideas and objects we have crafted as instruments—with anything noninstrumental. In the first category, agency is by reason of origin and purpose located in human beings. To do otherwise is to abdicate responsibility and hide our own purposefulness and agency behind the mask of doing "what technology wants," as if it were an independent agency. We have to hold those human beings directly involved in the finance, creation, and use of these instruments, directly, although nearly always collectively, responsible for consequences, even if—as is often the case—*they do not have full knowledge of what they are doing*. For others not so directly involved, responsibility in the sense of accountability may be indirect or nonexistent, although the consequences may evoke responsibility in the sense of obligation. Responsibility for consequences includes the way these instruments, once shaped by (some of) us, shape (so many of) us. In the second category, we as members of a common community share agency and unchosen responsibility in terms of obligation to further each other's ability to thrive.

There are category-crossing entities, and occupants of the middle ground, but that need not negate the categories. It is the denial of the possibility and actuality of a middle ground that is most dangerous. As Latour notes, because for moderns the middle ground between Nature and Culture was unthinkable, its inhabitants and activities remained obscured; therefore nothing could be bracketed and no combination ruled out. What is needed is our attentiveness to the categories and the crossings. Latour contrasts modern conceptual practice with that of nonmoderns (or those modernity classifies as "premodern"). "To put it

crudely: those (nonmoderns) who think the most about hybrids circum-
scribe them as much as possible, where as those who choose to ignore
them by insulating them from any dangerous consequences develop them
to the utmost." So much of this borderland is now being infiltrated and
patrolled primarily by instrumentalizing practices of knowledge making
and commerce that take for granted the notion of Locke that property is
"nature" "improved" to the useful.

Because the technoscientific enterprise we are about raises such
matters as collective agency and unchosen responsibility, it challenges
our classic definition of human moral agency. So to that subject, we now
directly turn.

Chapter Five

Reconsidering
Human Moral Agency

My discussion of moral agency thus far has centered primarily on the question of the moral agency of the more-than-human world, and in that pursuit I have indicated some of the challenges to conventional notions of who may be a moral agent when one takes a cross-cultural approach. But the concern for our characterization of otherkind is not the only reason to challenge our construction of the category of moral agency. [I also want to challenge our notions about ourselves as moral agents. And many of the previously raised questions have implications for the way in which those notions mislead us about the actual conditions and performance of human moral life.] Again I will bring to the fore voices that are not generally admitted to the conversation, from African American women and indigenous people, from religious voices that are not invested in (indeed they contest) dominant religion and secular philosophies.

But first, a short synopsis on what moral agency theory says and does. There are different frameworks within which the Western religious/philosophical has constructed its thought about moral agency. Aristotle and Aquinas focus on natural law and the goal of human flourishing. Jewish and Reformist Christian traditions have proceeded from the authority of the source of moral directives. Kant may have shifted the source from a divinity to a universal "moral law within" and emphasized human freedom without the Christian concept of sin, but the authority of principle is still ultimate. Charles Taylor refines that trajectory to the distinction of whether an agent merely assesses her or his actions based on their conformation to a set of standards, or whether s/he undertakes an assessment of the adequacy of the standards themselves. Moral agency is thus further qualified as "strong evaluation"—the capacity to form second-order

71

desires (the desire to desire something), to develop a language of evaluation, and to articulate the superiority of a choice.[1]

Despite these differences, the classical approach to moral agency theory in the West has two fundamental commonalities. First, what moral philosophers do when discussing moral agency is that they define qualifications to determine the condition of moral agency for the purpose of holding those agents accountable for their actions. For all their disagreements over which qualifications are appropriate, over priorities, emphases, and stringencies, qualifications are the focus, accountability the purpose. These include, minimally, the opportunity to discover and capacity to understand moral requirements, intentionality about one's actions, and the freedom to choose how and when to act. Second, for all the differences in understanding of human capacities and the differences that understanding makes to the condition of moral agency, there is a dominant idea of personhood that is nevertheless common to them.

Accustomed to the dominance of Western ideas of selfhood, agency, and community, we late-modern Westerners may not recognize what Clifford Geertz has noted, namely that "the Western conception of the persona as a bounded, unique, more or less integrated motivational and cognitive universe, a dynamic center of awareness, emotion, judgment, and action organized into a distinctive whole and set contrastively both against other such wholes and against its social and natural background is . . . a rather peculiar idea within the context of the world's cultures."[2] If our theories of agency are predicated on a peculiar concept of personhood, it behooves us to consider what implications differing ideas of personhood and different contexts may have for those theories. But this will not be a survey of thought regarding moral agency from the social margins, however worthy such an endeavor would be. What I have sought out are a few of those voices who, in challenging their own oppression and in making an honest assessment of their own moral lives, are revising the defining terms of the category in more inclusive, holisitic, and realistic ways for our understanding of moral life. I am particularly interested in those that may be helpful in understanding human moral agency as formed/malformed and integrated in a moral habitat.

This is not an exercise in replacing the modern Western conditions of moral agency operative in Euro-North American culture with terms derived from elsewhere. The descriptions of womanist and indigenous ideas of moral agency do not look familiar in either content or form. They are not category definitions, a list of qualifications to determine who is or is not a moral agent, or what acts are acts of moral agency. They are descriptions of human moral life as it is lived. They describe qualities, not qualifications, characteristics that help us to understand how people know the good and act on it. Narrative plays a pivotal role; it comes closest in method to a way of communicating about the thickness and complexity of moral life, a mode that exercises our moral imagination.

Further, I will not try to extract from this wisdom qualifications of moral agency in an attempt to reconstruct a conditional category definition of human moral agency. That is by nature an exercise in exclusion and can so easily become a pretext for oppression, as I will discuss in this chapter. While moral considerability is not necessarily a function of moral agency, women, racially constructed Others, the mentally challenged, children, and, as we have seen, nonhumans, have all at some time been excluded by conditions and assumptions about the capacity to fulfill those conditions, to the detriment of the respect toward bodily integrity, not to mention the ascription of rights. So I am wary of the conditional approach. Where conditions of moral agency are usefully set is in the strict determination of culpability, and even there this has its limitations, particularly in terms of determining corporate culpability. A conditional conception of moral agency beyond this application (to culpability) and a confinement of the concept of moral agency to the conditional is an inadequate approach to moral theory in two key areas: determining responsibility (as opposed to culpability) and understanding the performance of our moral lives. To name just some of the problems: If moral agency is located in "strong evaluation," then is the person who is not engaging in strong evaluation not a moral agent? Or not a moral agent at the moment, but potentially one because she has the capacity for doing so? What if, as is conceivable, we manufacture intelligent computers capable of abstract thought and even strong evaluations, perfectly capable of articulating these? Would they then be moral agents?[3] What about Murdoch's point that moral life is something that goes on continually, rather than being something switched on at moments of decision, that, indeed, it is all those moments in-between, the "background" of life, that is the more significant location of moral agency?

Having worked with the framework of ethos as moral habitat, and having considered portrayals of other-than-human moral agency and human moral agency in terms quite different from the assumptions of modernity, what can we say constructively about human moral agency? It is not the object here to dwell on the distinctiveness of human moral agency or to delineate which members of the other-than-human share which common traits with humans, and so could be accorded the status of moral agent. What I do here is glean insights from the portrayals given, both in this chapter and the previous ones, in order to begin to articulate a more adequate account of human moral agency.

In this chapter, we will consider the problem of human moral agency, meaning that we will investigate both certain ideas about human moral agency and the uses of those ideas as they are problems. I am hardly the first to notice the problematic nature of modern liberal constructions of moral agency. Critiques are made on the basis of the "postmodern" deconstruction of the subject, gender and race analyses, and communitarian philosophy.[4] I draw on some but not all of these. Despite those critiques, this complex of notions remains the operative one in Euro-North American society's practical moral reasoning and legal systems. Part of my objective here is to speak specifically to the power of this construct of moral agency in those contexts. It is not my ambition in this chapter to solve all the problems. I investigate, rather, these questions: What purposes are served by the category of moral agency? What problems exist, other than the alienation of humankind from otherkind, in our dominant constructions of human moral agency? What are the possibilities for reconceiving our understanding of humans as agents in a moral habitat?

Purposes and Problems of Moral Agency Theory

As noted above, the primary purpose of thinking about moral agency in the Western tradition has been to determine accountability, and to a lesser extent, praiseworthiness. An actor is deemed to be worthy of praise or blame according to (varyingly) his or her intent, object/means, or end. These are presumed to be subject to deliberate and free choice. The act needs to be uncoerced, with the agent possessing sufficient knowledge of the circumstances. The bases of evaluation and judgment in modern liberalism are usually assumed to be universal principles, in Kant's terms "the moral law," and the method by which they are reached a form of reason, usually following an established system of logic. But the cluster of notions represented by moral agency theory has developed over time and has historically served various purposes.

These purposes include providing a rationale and mechanism for exclusion and domination. It has been one of the main thrusts of this study to at least blur the line between humankind and otherkind when it comes to moral agency. This is, in part, because the claims to human difference in this regard have played such an oppressive role. Difference has been equated with superiority, as Anna L. Peterson reminds us. In this insistence on human distinctiveness, "the presumably unique qualities of humans, unlike the distinctive traits of other species, justify human domination" and not just domination of otherkind by humankind. "Oppressive readings of difference between humans and nonhumans are intimately tied, in history and in theory, to readings of difference among humans and to justifications of power and privilege based on difference. Especially, but not only, in the West, the definition of humanness in terms of a single quality has legitimized the oppression of individuals or groups assumed to lack that quality."[5]

This in and of itself would be reason enough to challenge the terms of moral agency, because moral agency and its component qualities as described above have been arguably the most resilient of the distinctive markers of humankind, where others, such as tool making and more recently, language, have not. Those who have suffered oppressive exclusion have taken up the question of moral agency and their own exclusion, and we will discuss some of their work below.

The construct of moral agency has also served as a basis for Western concepts of responsibilities and rights. (In some cases, moral considerability gets lumped in as well, on very shaky grounds.) The development of a theory of human rights is in its basic impulse egalitarian and liberating, but ironically it is precisely here that the moral agency theory has operated as a mechanism of exclusion and oppression, and its liberating work remains unfinished for many human beings. Exclusion from the category of moral agent has been at various times and places the rationale for denying people of color and the mentally handicapped even the most basic right of life, and they and many women and children have also been denied rights of safety, integrity, and self-determination. For example, although rights language was foundational to the national project of the United States, African Americans were not accorded the most basic rights to life and liberty. The various rationales included absurd arguments that these people were not fully human, lacking specific capacities deemed essential in modern constructs of moral agency, such as sufficient capacity for rational thought. Although suffering to a much lesser degree, white women were deprived of legal personhood on arguments of their lack of the same and consequent moral inferiority.

The designation of moral agency as an exclusively human quality is also repeatedly and vehemently used as an argument against "extending rights" to other-than-human nature. As DeGrazia points out, however, there is no clear connection between the specific qualifications of moral agency and any necessity to withhold the recognition of (particularly basic protective) rights. To paraphrase his argument, it is a great leap from the premise that all and only humans are moral agents based on, to use Geertz's words "a rather peculiar" notion of human personhood, to the conclusion that all and only humans are due full consideration or basic rights. In scrutinizing some of the supposed rationale, DeGrazia finds the Kantian "argument from dignity" consists not so much of reason as it does "rhetorical flourish."[6] Not that human beings lack dignity, but that there is no evidence that dignity necessarily depends on narrow definitions of moral agency alone, or that rights should be predicated on such a narrow idea of dignity. Dignity itself is a "birthright, a non-negotiable need,"[7] not subject to such qualification. Once again, the lines blur, and the identification of difference as a premise of value and power appears to be no more than an insistence on value and power in search of a premise.[8]

There is far more substance to the connection between constructions of moral agency and responsibility than there is with rights. This is true for responsibility either in the sense of culpability or in the sense of obligation. If one is capable of knowing and acting for her own or others' well-being, it makes sense that she bears a certain responsibility to do so. Some might argue that there is a lack of necessity here as well in certain modern liberal frameworks. There is no obligation to act for the good, merely sanction for doing harm. In the sense of culpability, there is a strong link between intentionally causing harm and accountability. And we regard the converse as equally true. We late-modern Westerners find repugnant to our sense of justice to hold other-than-human beings—or even human beings without the capacity to "know the good"—*morally* accountable (culpable) for their actions, even if the Bible in sundry places demonstrates no hesitancy in this.[9] And similarly, those who have not committed an act or did so under coercion should not be culpable. Those who did so *unintentionally* may be culpable, but the degree is less, and they are treated with greater leniency.

This strict qualification of culpability by the terms of moral agency—volition, intent, and knowledge—has been part of the Western story of liberation, and this idea thoroughly informs our legal systems. Those systems developed out of the particular social contexts of late medieval and modern Europe, and in those contexts, they are an achievement of moderation and containment of extreme injustices endemic to feudal hierarchies and powerful modern states. Western liberal models of moral agency enabled, for example, the rejection of vicarious punishment (a not uncommon practice of feudalism and slavery). Our ideas about moral agency contribute to the practice of presumption of innocence, certain standards for proof of guilt, and as we have seen, a measured sense of culpability based on degrees of volition, knowledge, and intent. These are achievements to value. Even here, however, there are serious gaps that arise out of the individualistic focus of moral agency theory as articulated above and in the way it is embodied in legal practices. These gaps can, in certain circumstances, become abysses that swallow the impulse for justice, reinforcing the existing power structures through the cultural constructs of those who hold power.

When it comes to individuals who clearly inflict harm voluntarily and with full knowledge that what they are doing is harmful or violates a law (or if it is a reasonable expectation that they should know, even if they do not), we do not have a difficulty sorting out culpability. If the act is done with forethought, with the opportunity for deliberative reflection and "strong evaluation," then the crime is judged to be even graver, and the penalties are commensurate. When harm done loses any degree of direct connection between individuals and acts, however, it becomes more problematic to sort out liability. In keeping with the spirit of our legal concern for protection of those not able to be considered culpable within the terms of our construct of moral agency, legal liabilities of a corporate nature are limited, or we resort to the legal fiction of corporate personhood, with liability limited by definition. It is interesting to note that in the modern West, a commercial organization, but not an animal or mountain, can have the status of person. Even those who argue for some sense of nondistributable collective responsibility remain in the position of arguing for that either against the individualist paradigm or in the same terms—choice, knowledge, intent, and actions deemed faulty according to some recognized standard. One of the problems here is, whose standards? Modern liberal agency theory accepts the idea that the "moral law" is absolute, fixed, and independent of context. But the standards appealed to here are actually those of the dominant society's ethos.

Modern liberal moral agency theories, with their emphasis on the singular agent, reinforce the assumptions of individualism. If the obstacles such individualism erects make it extremely difficult to think in terms of collective responsibility, those obstacles are even more intractable in matters of intergenerational responsibility, particularly with any sense of accountability. Our moral constructions appear most feeble when it comes to the question of answering for forebears whose intentions might have been deemed good according to the standards known to them at the time, within the ethos of their particular location, but which from another perspective or in retrospect are destructive and grossly unjust. Yet so many structural injustices and the harm from past injustices are multigenerational. If no agent or agents living can be judged responsible by the strict terms of liberal moral agency or if, in collective responsibility, the standards by which agents understood the moral law are themselves faulty,[10] is there then no accountability? Are biocide, genocide, and slavery simply to be forgotten when the generation of active perpetrators is past; is the slate wiped clean, even if suffering continues? Do we ourselves have no accountability for the consequences of our actions upon the well-being, indeed upon the very conditions for life itself, of future generations of the whole Earth community?

Finally, moral agency theory proposes to tell us something about who we are and how we live as moral beings; to use Kant's version, for example, which still holds great sway: we are free and rational beings who can determine our own wills in accordance with moral law. Within that broader purpose of describing human moral life, the language of moral agency offers a way to talk about claiming the power to act for the well-being of oneself and others. This brings me to the most neglected problematic area in the common modern construction of the dynamics of human moral agency: Its misrepresentation of human moral life as it is lived, a misrepresentation serious enough to lead us perilously astray in ethics. Consider this little piece of dialogue from a popular novel. The first speaker, Ishmael, has become a teacher to a largely disillusioned young man, the second speaker, who has had his youthful hopes of a transformed world lately and tentatively revived in the company of this teacher (who happens to be a gorilla).

> Ishmael thought for a moment. "Among the people of your culture, which want to destroy the world?"
>
> "Which *want* to destroy it? As far as I know, no one specifically *wants* to destroy the world."
>
> "And yet you do destroy it, each of you. Each of you contributes daily to the destruction of the world."
>
> "Yes, that's so."
>
> "Why don't you stop?"[11]

Why *don't* we[12] stop? Having the knowledge that our ways of life are exterminating species, wreaking climate havoc, fouling soil, air, and water; and knowing that bodies sicken and the stomachs of hungry children swell while the innocent are killed to provide oil and oil money for the wealthy—why don't we stop? Our constructs of moral agency tell us that, as free and rational beings, when we have such knowledge, we are capable of simply making a strong evaluation that sustainable and sustaining lives are preferable, deciding to change, and then doing so. Now we can also give many reasons—or in some cases more accurately excuses—why we don't, all of them true. To name a few: Those of us who have power, economic and political, are simply too comfortable; the destructive consequences of our actions are too distant in ecological location or time for us to be sufficiently motivated to change, even if we know that it would be the right thing to do. (That is, we may be moral agents, but we are not moral.) Or, we choose to close our eyes and pretend not to know what is going on, or that it isn't all that bad. We believe the story we've told of the place of the human in Earth, that the rest of life exists for our sake, that human needs are paramount. Or we keep ourselves and others passive and confused with the story that destruction of communities and inequity to the point of massive want among human beings is unfortunate, but secondary in importance to the "freedom" of markets.[13] We hope, against all evidence to the contrary, that we (or our children) will innovate the way out, without the need for sacrifice. Or, we cannot escape the structures of our lives just by choosing—how could we live outside the structures of cities and cars and mechanized agriculture? We're just trying to survive; it's too late to "go back." We are captive and individually much less powerful than we like to think; we can't escape that easily, even if we want to. And besides, if everyone "just stopped," we'd have economic and social chaos. It isn't that simple.

Exactly. And the dominant models of moral agency *are* too simple. They are, simply, wrong. It is not that human beings don't make "strong evaluations" or genuine choices based on moral principles; it is not that reason is nonsense or principles absurd. These models are false because they abstract individuals from moral habitats that are embodied, emotional, imaginative, *and more-than-human*, habitats in which we are formed/malformed and perform as moral agents. Nearly every "reason" given above for why we do not change our destructive behaviors points to the dynamics of an ethos, a network of norms and values embodied in and shaped and sustained by a whole range of practices. The fiction of the free and rational human agent keeps these practices as "background," "environment," diminishing our perception of their power and role in agency and thus distorting and handicapping our own agency. Most especially the participation of otherkind in these practices (including the degree of presence or absence and the quality of that participation) remains very "deep background," barely noticeable. We need to have models of human moral life that take notice, that are more realistic and more helpful in funding the wisdom, energy, and power to act for the flourishing of Earth.

Again, the rest of this chapter is *not* an attempt to replace one abstract, universalized notion of humanity or moral agency with another or to appropriate the self-definitions of others. The sources here overtly insist on their integrity within particular ecological locations. This is affirmed in the concept of ethos as moral habitat. That does not mean, however, that we cannot learn from one another or do not at times need to be held to account by one another, across ethē. An ethos is not a self-contained moral living space but one related, as intimately connected to others, even if as apparently distant, as the rainforest is to the tundra. Not all the problems presented by the false tale of human agency analyzed above are addressed, necessarily, in this approach. But some are.

Human Agency in a Moral Habitat: Womanists

The completely free and rational agent is a false portrayal of human moral agency, even for those in positions of privilege who hold it most firmly and for whom it provides various shields and benefits. Because of those shields and benefits, the place to turn for more realism is to those for whom it has not been so beneficial, to those who have more clarity because the falsity of this fiction is a lived, sometimes painful reality. Therefore they already identify their moral agency in different terms. This move is especially necessary in order to remain on guard against rein-scribing oppression in the process of reconstructing human moral agency in terms of "moral habitat," since social relations and their networks of norms and values can, as feminists like Chris J. Cuomo remind us, produce and reproduce "prejudice as well as connectedness, oppression as well as resistance, confining norms as well as identity."[14] We will continue to discuss this in the next chapter, on doing ethics in a moral habitat, but it is inescapable when talking about human agency. This is made abundantly clear in the work of Katie G. Cannon.

Cannon insists that a situation of unelected suffering, struggle, and restricted freedom demonstrably changes the ground of moral life and negates the dominant assumptions of ethics. The differences from dominant systems encompass not only the values and norms themselves, but also the processes of moral formation, the bases and modes of ethical deliberation, and the qualities and exercise of moral agency. "Black women live out a moral wisdom in their real-lived context that does not appeal to the fixed rules or absolute principles of the white-oriented, male-structured society. Black women's analysis and appraisal of what is right or wrong and good or bad develops out of the various coping mechanisms related to the conditions of their own cultural circumstances. In the face of this, Black women have justly regarded survival against tyrannical systems of triple oppression as a true sphere of moral life."[15]

Cannon argues that the literary tradition of African American women, "the nexus between the real-lived texture of Black life and the oral-aural cultural values" constitute both a moral source and moral practice. In her analysis of the life and work of Zora Neale Hurston as an "especially concrete frame of reference for understanding the Black woman as moral agent,"[16] she identifies womanist moral agency as characterized by "invisible dignity," "quiet grace," and "unshouted courage." The first, invisible dignity, she associates with Alice Walker's term "unctuousness," a smoothness and slipperiness that enables an agent to negotiate situations of insincerity and even of danger—with "her head in the lion's mouth," being forced to "treat the lion very gently."[17] In describing this quality, Cannon cites Hurston's comparison with the dignity of the earth, capable of "soaking up urine and perfume with the same indifference."[18] With this "invisible dignity," an agent can assess her situation for herself, demythologize "whole bodies of so-called social legitimacy," and maintain a balance of forethought with "discerning deliberation." In a context of limited choice and imminent threat, this realistic moral agent does not make perfectly free choices according to a system based on abstract principles but on her own canny discernment of the genuine choices available in the midst of living in the circle of life.[19] Choice is an extremely significant exercise of moral agency even if it does not change the circumstances.[20]

Yet agency is not fully expressed by choice where that is simply selecting among preexisting alternatives available. There is an active world-shaping dynamic in womanist moral agency, a creation of "something new where nothing was before"[21] and destruction as well: "The Black community is called forth to fashion a set of values in their own terms as well as mastering, radicalizing and sometimes destroying pervasive, negative orientations. For Hurston, this is the essence of quiet grace."[22] This characteristic of quiet grace is forged in the context of living in close association in circumstances that do not provide "the protective privileges that allow one to become immobilized by fear and rage,"[23] for life is at stake. It is also a collective creation over time. According to Cannon, it is important to understanding moral agency in Zora Neale Hurston's fiction that "the protagonists are always connected with the general history of the race in the context of community. . . . [E]ach generation is dependent upon the last for its understanding of moral wisdom and in turn, each new generation creates it for the next."[24]

The tight reciprocal associations of responsibility and free choice predicted by the dominant liberal models of moral agency simply do not describe moral life adequately here. There are intergenerational dimensions and situations in which choices made do not effect any change in circumstances. Yet the condition is not one of irresponsibility but "forced responsibility," out of which evolves the third characteristic of womanist moral agency identified by Cannon, "unshouted courage."[25] This is not courage according to the dominant criteria, which assume a range of choices and sanctions as prerequisites for moral responsibility, and consider courage "a virtue only when it is distinguished from spurious, physical fear. . . . Black people live, work, and have their being within less gracious boundaries."[26] They develop as a communal attitude a moral element of courage "annexed with the will to live and the dread of greater perpetrations of evil acts against them."[27] Here is grit and long-term energy for the struggle. Unshouted courage is not "grin and bear it." Much more than that, it "involves the ability to 'hold on to life' against major oppositions. It is the incentive to facilitate change, to chip away the oppressive structures, bit by bit, to celebrate and rename their experiences in empowering ways."[28]

In her exploration of the relation of the Black church and womanist moral agency, Cannon also points to the mystical, spiritual grounding of agency in the writing of Howard Thurman. She notes that for Thurman, "No matter how restricted moral agents might be, the experiential-mystical element calls on each person to act and reflect the divine in her actions. . . . A committed spiritual life is necessary for an accurate sense of moral agency."[29] In his words, "In the moment of (mystical) vision there is a sense of community—a *unity with all life*, particularly with human life" (emphasis mine).[30]

In her discussion of womanist moral agency, Cannon is not particularly explicit in delineating its relation with that "deep background" of otherkind. She does not develop the ecological elements of Thurman's mysticism in her discussion. For example, in relating an experience of his youth in Florida, Thurman describes, "I had the sense that all things, the sand, the sea, the stars, the night, and I were one lung through which all of life breathed. Not only was I aware of a vast rhythm enveloping all, but I was part of it and it was part of me."[31] Cannon provides glimpses of the "deep background" at work in African American ethos in the brief reference to the earth as a model for Hurston's "invisible dignity" and in a few other mentions of Hurston's work, but that is not a primary focus for Cannon. This is not the case for Karen Baker-Fletcher, who selects this passage from Hurston to open her third chapter, "Why the Hurricane?": "They huddled closer and stared at the door. They just didn't use another part of their bodies, and they didn't look at anything but the door. The time was past for asking the white folks what to look for through that door. Six eyes were questioning *God*."[32]

With this, Baker-Fletcher introduces the idea that experiencing the power of God in connection with the rest of nature is not a simple, tranquil, comforting affair. "Sometimes nature's beauty is fierce, its power frightening, because it reminds us that it is larger than our small bodies and the puny homes we build to shelter ourselves. At such times, one feels less ecstasy and more awe, even terror."[33] According to Baker-Fletcher, Hurston is hearkening back to ancestral understandings here, understandings of respect for nature and "elemental understandings of God as moving in wind, rain, sky, and earth."[34] (Such understandings are, of course, completely coherent with the Bible as well.) The experience of vulnerability before such immense "strength brings us face to face with the power of our own selves and with our own mortality. It reminds us of the importance of respect and restraint in our daily lives." There is an important difference between Baker-Fletcher's appreciation of that strength (within limits) and the conflictual stand against "nature" that is so prevalent in Western inheritance, particularly after the plagues. This biblically informed, ancestor-honoring womanist links humanity *with* the strength and power of this God-created and -activated world, not against it in a structural relationship of enmity. "We, too, are nature, we who according to Genesis are dust of the earth, we who scientifically are known to consist of earth, water, oxygen—the elements of this planet. We are also a force of nature. Like the rest of creation, we have the power to inspire comfort or terror in those around us." Such strength comes with a moral imperative, as well as motive power. "What a violent world we risk producing if we do not gain control of our own power to destroy and create, if we do not center ourselves in the creative spirit of life itself, God, who is love. . . . Weak enough to be overcome by our environments, strong enough for our carelessness and potential for violence to destroy one another and our planet, let us choose what kind of force of nature we will be. May it be positive, wholistic, and salvific for all."[35]

Baker-Fletcher evokes the sustaining power of touching Earth, but does so without flinching from the complexity of living in afflicted places. Contaminated soil in urban communities and in rural communities of color used as toxic dumping grounds prevent the uncomplicated "planting with praying" that traditionally connects "dust with spirit" and allows us to touch something "deep in our own spirits and the Spirit of God."[36] She names the absence of healthy encounters with more-than-humankind in the lives of many urban young people as injustice. "Justice is absent in the lives of many young black women and men who live in a world of steel and concrete."[37] Freedom includes the land. "As long as we can love the earth and eat from its loving, nurturing fruits, we are free. . . . As long as we can walk the land, we are free. But when our land is poisoned, we are not as free."[38]

Seeing Indiana Avenue in the 1970s, at the height of its devastation, brings home to Baker-Fletcher the connections of spirit and dust to the devastation of a blighted urban area, as well as the illusion of her own situation of relative improvement. "When my feet pound the concrete I feel the rhythms of sorrow, rage, and keepin'-on laughter. I knew then and now that my own economic paradise is no real paradise. I knew that what we have lost as a people is deeper than money and privilege. It is God, land, and the freedom to claim both in our own way."[39]

Baker-Fletcher locates human moral agency, as Cannon does, in community that transcends temporal boundaries and embraces the spiritual. The voices of preachers and grandparents, parents, aunts, and uncles admonishing young folk to "remember where you came from" point to this source of moral authority and agency. Baker-Fletcher recalls this being put in a wider context in a memorable sermon by Pastor Covington, who pointed out that traditionally this is not just a human community: "Seeming to look each member of the congregation in the eye, he said: 'We need to remember who we are and whose we are. We are people of the land and we belong to God. We used to know that.'" By locating human well-being and agency in community so broadly conceived, Baker-Fletcher unites God and Earth as both directing our action and providing motive power: "God loves the earth fully. By loving one another and every sentient being—even the rocks who cry out—we love God. In this love we are called to resist the poisoning of peoples and the earth. . . . We must awake from the seductive sleep of spiritual torpor and remember who we are."[40]

My exploration here, once again, is not a search for "new universals" with which to draft a "generic" definition of human moral agency. "Who we are" is a matter of our common being as human beings, and our membership in a more-than-human community, but it also is located in the particularity of history, social, and geographic locations. I will explore this further below, but I would like at this point simply to make a few observations about womanist articulations of moral agency. These can be suggestive for reconceiving human moral agency in other ecological locations, even though some aspects may be unique achievements, made under the oppressive and unjust circumstances of African American women.

Cannon, Baker-Fletcher, and other womanists describe moral agency that develops both within and in resistance to a situation of triple oppression. Its particularities are not accidental but essential to the understanding of womanist moral agency, and more importantly to its formation and exercise. I had written above that what we need is realistic models of moral life that fund the wisdom, energy, and power to act for flourishing. But in womanist contexts, the use of the word "survival," rather than flourishing, is emblematic of the difference that limited freedom makes. Agency here is called into being and activated in the work against exclusion and oppression, not, as we saw with Western definitions, delineated as a form of exclusion and oppression. "Forced responsibility" is not bound by the tight ropes that anchor the moral dimensions of individual culpability, but let obligation slip, fray, and snap when it comes to collective and intergenerational responsibilty. Responsibility unchosen calls forth the determination of "unshouted courage," a fund of strength that comes from an enduring, not a frenzied energy. "Discerning deliberation" speaks of wisdom that understands the genuine choices available in a less than ideal world, but I think my terms of "wisdom, energy, and power" ignored the presence of "invisible dignity" to which this kind of wisdom belongs.

Finally, community is also essential, not incidental, in the self-understanding of womanists regarding their moral agency. Quiet grace is forged through living in close relationship with others. Community, "the arena of moral agency," is based on love.[41] This community is intergenerational; past suffering and fortitude are a source of moral grounding, and moral counsel is "implicitly passed on and received from one generation of Black women to the next."[42] To some extent, this community is acknowledged as more-than-human, in a way that sustains (or in deprivation wounds) moral agency, and in a way that challenges us to take seriously other-than-humankind as moral agents and intimate, necessary allies. Community here participates the fullness of personality, which in Thurman's vision is larger than mere individuality, but doesn't negate it. Cannon quotes him as saying that personality "is a fulfillment of the logic of individuality in community."[43]

In contrast to this vision, individuality and community are often framed in Eurocentric discourse in either/or, or even hostile, terms. Either one holds to the vision of a relationally constituted self and relationally performing moral agent or to that of the hyperindividualistic, autonomous, self-authorizing agent of Kant. To consider how to come at this differently, we turn to the ways in which the concept of personhood in indigenous North American nature-cultures, introduced in an earlier chapter, bears on issues related to human moral agency.

Human Agency in a Moral Habitat: Indigenous North American

While there are areas in which the indigenous and Eurocentric concepts of autonomy overlap, there is interesting and significant divergence relevant to moral agency. There is always great difficulty in translating across cultures. "Autonomy" is not a word I would have chosen to describe aboriginal concepts, because the peculiar ideas of personhood in our culture are wrapped up with Kantian notions of autonomy, but "autonomy" is indeed the choice of many indigenous people in communicating their self-understanding. Once again, this may be an example of the "distinct Englishes" revealed in Valentine's work with indigenous language practices mentioned previously. The task here is to attend to the fullness of what is being said in the Native context with the use of this word.[44]

As briefly stated in a previous chapter, personhood in Native nature-cultures assumes a person is a self capable of directing his or her own behavior, and who is due respect, so that others are not entitled to infringe upon a person's sense of her/himself, or to manipulate or coerce a person. A person is to be treated as an end, never as means.[45] So far, this fits with Kant. But as we discovered, this is not an exclusively human state, but is shared by and constituted in "all my relations." "When we have our sacred ceremonies, like the sweat lodge, we end it by saying, *No'kamaq*, which means 'all my relations.'"[46] This includes animals, plants, insects, spirits, elements, and cosmic or geological features (stars, mountains, rivers), as well as humans, in webs centered on an extended family and the constructions of particular animal and plant affiliations with individuals and families.

Indigenous autonomy is not acquired after an extended developmental period of heteronomy, in which a child's body and activities are considered rightly controlled by adults in authority. In most Native worldviews, autonomy is recognized as belonging to children at a very young age, when a child begins to talk or walk.[47] It is not predicated on any capacity for abstract thought, or on being able to grasp and assent to principles. Autonomy is inherently part of the definition of personhood, and as we saw in the previous chapter, this is not an exclusively human category. We have definitely left Kant, and most of Western thought, behind here.

Indigenous thought moves beyond any simplistic opposition of the autonomous versus the relational self. Personhood, with its autonomy, is granted; simultaneously, the self is constituted in one's particular relations, including more-than-human relations. These relations entail responsibilities as well as identity. Indigenous concepts of selfhood are as much about responsibility as they are about autonomy; the two are inseparable. Responsibility, like autonomy, is something not acquired by actions or assent. According to Gwen Bear Orechia, a Maliseet, "You have a responsibility to be balanced, to have your community balanced. An Indian is born with responsibilities which *must* be taken on."[48] These responsibilities are learned, but not chosen. They are given, inescapable. As Susanne E. Miskimmin reports about another Algonquian group, "the Nishnawbek are *responsible* (emphasis original) for living in harmony with all parts of Creation. Many First Nations societies understand themselves as a complex of responsibilities and duties. These responsibilities are deeply felt and are *crucial to the very definition of what it means to be human*"[49] (my emphasis).

Responsibility here means being responsible for the well-being of others. Whereas autonomy that is not integrated with this kind of responsibility may go no further than an imperative of acting in ways that do not interfere with others, the integration of the two here creates an *obligation* to promote others' welfare actively, but *without violating their own autonomy*. Accountability is likely to be expressed in personal terms, rather than in terms of action or condition—Who (*sic*) do you answer to?[50] rather than Under what conditions is one answerable for something? Responsibility in terms of being responsible *to* is also not temporally bounded, but includes responsibility to ancestors[51] and future generations.[52]

The emphasis on the personal, being responsible *to*, means that autonomy and responsibility are also thoroughly embedded in community. They are understood in terms of development of competence and self-sufficiency in order to be able to contribute to the community and to avoid being a burden on others.[53] One is simply not an autonomous agent in the absence of or over-against community—and again, the community includes other-than-human persons.[54] Community is not chosen, but is established by kinship, and kinship terms are used for nonhuman relations as well as human.

An individual is expected to assume responsibility to the community for consequences of his or her behavior as well, but this is where Western law (and concept of autonomy) is perhaps most significantly different from indigenous traditions, in that the individual and particular acts are not abstracted from the community. Rupert Ross, in his study of indigenous justice practices, draws some contrasts with the dominant legal framework. "Western law seems to assume that we are captains of our own ships and that each of us is equally capable of moving out of antisocial behaviour on our own, just by deciding to do so. Traditional wisdom suggests that each of us rides a multitude of waves, some stretching back centuries, which we cannot fundamentally change and which will still confront us tomorrow." And he says, "according to Western law, 'taking responsibility for your act' means little more than acknowledging the particulars of the illegal act, then paying a proportionate price in punishment. Traditional wisdom suggests that acts are important only for their consequences on the mental, emotional, spiritual and physical health of all those affected, including all the *offender's* relationships." Taking responsibility for an act involves "coming to feel some portion of the pain" that has been caused to everyone.[55]

For understandings of moral agency in terms of finding wisdom and power to act for healing, for balance, for the struggle to reorder the colonialist structures that oppress Native people, I turn to the story—told by a group of Maliseet women—about their own exercise of moral agency (although they do not use the term). These women led a campaign to overturn one hundred years of legislated sexual discrimination against First Nations women. Through the Indian Act, the white government had taken it upon itself to define who is, and who is not, entitled to the legal status of "Indian." With that status come certain benefits, including such basic things as housing on reserves. The act paid no attention to the lineal descent patterns of indigenous cultures but arbitrarily decided according to Eurocentric patriarchal bias that Native women who married white men were no longer "Indian," while white women who married Native men, were suddenly "full status" Indians. Like African American women, Native women suffer triple oppression. Their struggle was not only with the Canadian government, but also with men in their own communities (some of whom had white wives) who were in positions of power in the patriarchal structures imposed on those communities by the Indian Act. In speaking of their struggle, these Native women identified the practices that constituted their own moral agency. They claimed themselves as authentic sources of knowledge and their own analysis as truth: "We all knew that no government agency—be it white or be it Indian—was going to tell us we were no longer Indian, when we *know* we are Indians. . . . I knew I was Indian—nobody took the Indian blood out of me. Therefore I think all of us women decided there's this whole discrimination thing going on, and it's all geared against the *women*. . . . When I look back I see that we became more and more aware of the Indian Act standing behind a lot of our problems" (Mavis Goeres).[56] That knowledge consisted of identity, conviction regarding the injustice of their situation, and the recovery of their own heritage, another important element in their exercise of moral agency.

Strong traditions of relational community enabled them to find strength and endurance in each other. "A lot of us had blistered feet, but we went on. There was (*sic*) many times we'd be so discouraged, some would want to turn back—and some did—but most of us all kept each other going" (Mavis Goeres).[57] And once again, the community that participates in moral agency is more-than-human. "What Indians need to know is this tie that we have first to the family and then to the earth" (Juanita Perley).[58] Perley and other Native women identify spirituality (both Christian and traditional) as important in their agency, as affirmed in the traditional refrain, repeated often in ceremonial practices, "all my relations."

This concept of community as something not constituted by voluntary affiliation or accident of location (although deeply informed by place). It is community that includes ancestors and future generations and is recognized as having more-than-human members. This takes the indigenous sense of self-in-community beyond that of even most Western communitarians. But it is really not all that rare or strange in the world or through history; it is the idea of an abstracted, self-contained whole "set contrastively both against other such wholes and against its social and natural background" that is "rather peculiar."

Reconceiving Human Moral Agency

While each of the sources we have looked at in this chapter—womanist and indigenous Algonquian—belong to distinct moral habitats, there are a number of common elements. One way to proceed with reconstructing our ideas about human moral agency is to identify those not as conditions but descriptors, a basis of inductive reasoning.

Both womanists and Algonquian people claim "canny knowing" as a source of moral authority, a knowing that is not subject to narrow epistemologies, definitions of rationality, or specifications of strong evaluation. They insist on particularity of location, including the location within a multigenerational community, as a parameter of moral agency. This community is claimed as a source of identity, moral wisdom, endurance, and power that make moral agency possible—but also as making claims of unchosen responsibility that expand the reference of moral agency beyond the self and the consequences of individual choice. Both womanist and indigenous voices present this community as more-than-human and point to spiritual practices and traditions as integral to moral agency.

One thread through all of this is that one is not a moral agent by oneself. Moral agency is not a matter of isolated individual acts. It is not even a matter of making judgments and acting according to some collectively woven set of norms or the cumulative action of a collection of individuals. It is dynamically interactive; it is shared. Actions shape norms, not only follow them. Our agency is activated by that of others. Your agency and mine are bound up together. Oppression calls forth agency in the oppressed and makes a claim on the agency of the oppressors. If this claim is not answered, the moral agency of oppressors is corroded. Forms of taking responsibility may involve "coming to feel some part of the pain of another." It is not only about pain, however, for in the avoidance of pain we also lose genuine joy, authentic wisdom, and strength. This is true in the context of all our relations, other-than-human as well as interhuman. The implication is that what is other-than-human shares in and is necessary for any form of moral agency.

This study can only be the beginning of the process of reconceiving human moral agency, working toward a more adequate description. But if this description does have some validity as a depiction of human experience across moral habitats, there should be evidence in other cultures as well. We could further this process by seeking elsewhere for testimony to human moral agency as something that is activated by and in community, including other-than-human. This is not difficult. The Bible also grounds human moral agency in the relationality of community spanning multiple generations, without any diminution of the worth of the individual human being or diminution of personal responsibility. The human made in the image of God is not set apart but immediately related as God is related (and as we have seen, this relationship is conversational and collaborative) with the earth and all other creatures. Elsewhere, the substance of the human is that of the arable soil itself (Gen. 2:7) and "intricately woven in the depths of the earth" (Ps. 139:15). As Michael Welker observes, "It is God the creator who constitutes the complex unity of the person and the complex interdependence of the creatures."[59] Dietrich Bonhoeffer ties scripture's positioning of the human as moral agent to a strand from another natural-cultural fabric altogether, the story of the powerful giant Antaeus, who could not be overcome until he was lifted from the ground. His strength had come from the earth. "The person who would abandon the earth, who would leave its present distress, loses the power which still holds him by eternal, mysterious forces. The earth remains our mother, just as God remains our Father, and our mother will only lay in the Father's arms those who remain true to her."[60]

What this suggests is that not only does Earth and its distress call our moral agency into being, we are allies in responding to the creative power of God. We need the other-than-human Others. We need to regard otherkind as moral agents, actors for their own and others' well-being, and to deal with them in this spirit of mutuality. It is not a matter of our "extending" agency to them in a paternalistic fashion, or looking out for their interests from a sense of dry-as-dust duty. And we need this because we ourselves are not moral agents apart from Earth. Without life in community in its fullness, when we set ourselves apart and pretend we are suspended above the earth, like Antaeus, we lose our power. We lose the wisdom and power to act for the flourishing of life, for the well-being of ourselves and others. We cannot stop the destruction of Earth based only on our own knowledge and will. Without listening to all the voices of creation, we lack knowledge of how to be and act. This is not a matter of merely discerning a codified set of laws, but an ongoing conversation. As we listen, not only speak, we learn and we change. We are shaped imaginatively and morally in and by the whole company of Earth. We can choose to act *with* otherkind in love, justice, and power, or against them and apart from them—but then ultimately only in destruction, ours as well as theirs. That is some of what it means to be a human agent in a moral habitat.

Chapter Six

Doing Ethics in a
Moral Habitat

I had barely begun writing this book on the morning of September 11, 2001. The events of that morning are perhaps as close as the human race has come to an experience shared around the world almost simultaneously. But the actual experience of people varied greatly, depending very much on who and where you were. I happened to be in Manhattan, in Morningside Heights, discussing this work with my mentor, Larry Rasmussen. We were far enough away from the World Trade Center not to realize at the moment the magnitude of what was happening at the other end of the island, although foreboding grew as our conversation was halted repeatedly by the sounds of sirens screaming out of the firehouses of the Upper West Side—they had a straight line down the West Side Highway to the World Trade Center. The least of my memories of the hours and days following come from the well-known images from television. It is the sounds of those sirens, of fighter jets, helicopters, and an eerie silence so unlike New York that I remember. It is the sight and smell of smoke. It is the touch of hands clasping and voices whispering and raised in grief, relief, and anger, and in songs and prayers—in English and Spanish, in Hebrew and Arabic. I remember tears and candles and photographs. I remember the taste of shared meals, without bread, which was quickly in short supply once the island was sealed off. Communion of another sort, perhaps, but real communion nonetheless. I remember fear, certainly not what I would call terror, but paralyzing uncertainty. My greatest fear was of what the United States would do and become. I wish I could say that fear was unfounded.

At first it seemed that events had overtaken some of the main issues I was pursuing, particularly the question of moral agency of otherkind, and rendered them superfluous. On further reflection, I would argue that not only is reconceiving moral agency—of both humankind and otherkind—a vital step toward the transformations necessary to have a future worth living in, but there are also conceptual and methodological moves in this cross-cultural, interreligious study of ethics that are part of a promising development of ways of seeking deep understanding across difference, ways that attempt to leave behind ethnocentric, colonialist habits. Such attempts are necessary to the flourishing, indeed the very survival, of life in the world in which we now find ourselves.

What I have done is to start in a different place from most discussions of moral agency or even cross-cultural ethics. I start with Earth. Earth, one Earth, at once a whole and incomprehensibly diverse. I theorized that local topography and biota participate in (though they do not determine) many of the elements which are woven together in the construction of a set of cultural values and norms, its landscape of moral imagination. This is one aspect of "moral habitat." The other aspect is to examine different constructions of the capacities of otherkind to know and do the good in Mi'kmaq and other Algonquian traditions, in the Hebrew Bible, and in post-Cartesian science. For the Mi'kmaq, we live in a community of relatives. For the Priestly writers, "all flesh," clearly inclusive of more than humankind, are both agents of and respondents to the will of God, and the other-than-human as capable as humankind of deviating from it. These conceptions, along with evidence gathered in the course of science in light of evolution, ethology, and the first meaning of "moral habitat," make it possible to conceive of human moral capacities as not only emergent from but shared with the rest of the biotic community.

From there, we can already put into question the composite construction of moral agency as received from the Western traditions. But I also raised further questions of its adequacy in terms of realistic portrayal of *human* moral life, corporate culpability, and intergenerational responsibilities. To help reach a fuller understanding of human moral agency, I incorporated the concept of moral habitat, in both its aspects, with the wisdom of womanists and indigenous ideas. So where do we go from here? How might we do ethics in a moral habitat? First, I will discuss the general directions indicated for method by the concept of ethos as moral habitat and the implications of a fuller sense of moral agency for our participation in the community and future of Earth. Then I will demonstrate the ways in which the ideas discussed in previous chapters open up particular ethical discussions through three examples: intergenerational responsibility, abortion, and genetic engineering.

Method in Ethics: Moral Habitat

The concept of ethos as moral habitat, analogous to a physical habitat, broadens the scope of the *means* of doing ethics. As explained in the first chapter, the practices that weave the networks of norms and values that make up an ethos, and the reflection upon them that constitutes ethics, are not restricted to disembodied cognitive practices, as so much ethical reasoning in the modern West understands itself. Ethos formation and reflection take place in and through myriad cultural practices. To recall the list: Language; stories, from sacred myth to entertainment to gossip; arrangements of space (architecture, landscape) and time (calendar, festival); ritual and ceremony; contemplative practices; procurement and preparation of food and what foods are eaten or forbidden; conduct of bodies, clothing, gestures, and attitudes toward bodies; gender and sexuality; practices of healing and reconciliation; power relations; inheritance, ownership, and trade; practices of knowledge acquisition, verification, and transmission; procreation and childrearing; poetry, music, craft, dance, and imagery; games and other play; including the technologies employed at times in much of the above. These practices are not merely expressions of or means of expressing norms and values; they shape moral imagination and norms and values. The other-than-human, physical habitat is so involved in these practices that "social construction" is too narrow a term to understand a process that is natural-cultural.

The practice of constricting our concept of doing ethics to a second-order disembodied cognitive practice, such as the West has conceived logic, is to assume that all other cultures do ethics in the same way, and if they do not, they are not doing ethics at all. This is absurd. Secondly, understanding the thickness of ethos, its contextuality and the embodied nature of moral deliberation, challenges the Rawlsian proposition that there can be anything like an "original position" in which sex, nationality, race, religion, and so on, are suspended, and only speculatively entertained, for the purpose of ethical discourse.[1] Subjects can imaginatively enter the experience of a differently situated other, but only to a limited degree, and most effectively through multiple avenues of practices. Doing ethics in a moral habitat is decidedly an entangled and messy business. I would argue that is nevertheless a more useful approach if only because it more accurately engages human moral life as it is lived.

Moral life is inadequately portrayed by a preoccupation with decision making. Ethos as moral habitat sets the process of moral decision making (usually in a situation of competing norms) that occupies much attention in ethics, within a much larger scope. The human moral life is inadequately conceived—as it has been particularly since the Enlightenment and in some form throughout Western philosophical history as a whole—as the reasoned arrival at particular principles and the will to adhere to them. *Most* of human moral life—that is, our activity in relationships expressing estimations of what is valuable, right or good, and how to behave—lies between, around, and beyond these decision-making moments, as Iris Murdoch described. Human moral life takes place within the boundaries of an ethos, without the imaginative concepts, principles, or values of the ethos necessarily being subject to explicit articulation or intervening conscious reflection. Moments of decision or choice may reveal the underlying network of norms and its structure, and principles may be abstracted from the embodied ethos, the practices of a community. But principles are derived from experience in embodied community life rather than premeditated and then implemented. The network and its structure preexist and shape life even when we are not aware of choice making, and when choice is precluded. This does not mean we don't use reason and principles to evaluate our actions. I'm using them now to evaluate our constructs of thought, but I am doing so with awareness of the pedigrees of reason and particular principles, and their limitations.

Doing ethics in a moral habitat does not make moral dilemmas or choices less real or agonizing, but reveals the parameters around them and widens the scope of understanding moral agency: People still nurture and care for one another, sacrifice their own interests on behalf of another, and act for their own and the common good, without necessarily deliberating about it. Such deliberation is not superfluous, but it is derivative in moral life; some form of ethos is the required matrix in order for ethical discourse—regarding principles or competing goods—to make any sense at all. "Moral habitat" holds together all of these aspects of ethos—its formation and communication in embodied practices with their felt as well as understood values, meanings, and purposes—with the ecological understanding of habitat as the larger biotic community, and it does so with the economy of metaphor.

I don't think of "ecological ethics" as one among other special concerns of "applied ethics," a specialty akin to business ethics, sexual ethics, or medical ethics, one that concerns itself with "environmental issues." By tracing all the practices of nature-cultures as productions and producers of the network of norms, meanings, and values that constitute an ethos—and at the same time insisting on the *natural*-cultural origin and consequences of these norms and practices—it becomes obvious that the practices of commerce, sex, healing, and so on, are inextricably intertwined, with each other and with the other-than-human members of the earth community.

Let's take medicine, for example. Looking at medical ethics in the context of a moral habitat raises the awareness that practices of medicine are linked to pathogens and movements of people, to sex, capitalism, politics, knowledge-production, war, technologies, mining, food production, water use and availability, waste management, energy use, plant evolution, constructions of intellectual property and patent legislation, colonialism, food, entertainment, gender practices, and so on. As an illustration, let me contrast the conventional approach with that of "moral habitat." The conventional approach to type 2 diabetes might discuss the patient's choice in diet, exercise, and pharmaceuticals, or the ethics of clinical trials. A moral habitat approach would look at social and economic structures, at the relative positions of people due to gender and race, at agri-food industry practices. It may be very well to tell a poor woman in North Manhattan to eat a healthier diet and get exercise, but desirable food may not be within her means or geographic reach, and the air she breathes may make it much more difficult for her to exercise, due to truck and automobile traffic, and the location of bus depots without regard to topography and airflow. She is unlikely to have easy access to safe physical recreation. To shift from the particular to the general in the discussion of medical ethics, we need to look at not only who is covered for what procedures in medicine, but the cultural drive to prolong life at often extreme costs. This is not merely the inheritance of the Hippocratic philosophy, but a product of natural-cultural history in the plagues, in the Western abhorrence of decay, and its fear of the natural world, especially the inescapability of death and the necessity of death to the sustenance of life. Doing ecological ethics involves holding all this together. It demands "ecological thinking," reweaving the foreground into the "background" that is usually ignored when we extract ethical questions and concerns away from the "habitat" in which they live. That is ecological location in a fuller sense, with a slightly expanded idea of "situated ecological ethics." But there is still more.

One thing I am trying to accomplish with the metaphor of moral habitat, developed in this fashion, is to find a means of *orientation* and *navigation* within an ethos. I am constructing ways for the person "native" to a moral habitat to articulate its topography in an explicit way that doesn't demand she merge her identity or worldview into that of a "generic human." I seek out ways for an "alien" to discern the contours of unfamiliar territory, if not with the bone-marrow knowledge of the "native," then with at least a modicum of awareness that she is "not in Kansas any more." That ethics *is* a contextual endeavor is hardly a new idea, but discernment of context is not as straightforward as might be assumed. There is more to location than "a listing of adjectives or assigning of labels such as race, sex, and class," says Donna Haraway. "Location is not the concrete to the abstract of decontextualization. Location is the always partial, always finite, always fraught play of foreground and background, text and context, that constitutes critical inquiry. Above all, location is not self-evident or transparent."[2]

This work is an attempt to approach cross-cultural ethical discussions in a holistic way, without the excessive privileging of a gendered eurocentric method and assumptions. One of those is that there is a "view from nowhere" and another is the category definition of moral agency. I am trying also to find paths of navigation between the proverbial rock and hard place, or perhaps more aptly between the rock and the quicksand, of moral absolutism and total, or as Midgley calls it, "simple" moral relativism.[3] As she observes, nature-cultures are not isolated, watertight containers; while cultures differ, "they differ in a way which is much more like that of climatic regions or ecosystems than it is like the frontiers drawn with a pen between nation states. They shade into one another."[4] To put the focus on practices as they constitute a moral habitat and their common formation within an enlarged sense of Earth community acknowledges both the integrity of differences related to location *and* the common origins of our moral capacities and sensibilities, as well as our ongoing shared dependence on each other—human and other-than-human—for the well-being of all. This is an aspect of moral habitat that has been taken up here in only an initial way, in the cross-cultural approaches to other-than-human and human agency. There is much room for further exploration.

One of the most promising avenues for cross-cultural and interreligious explorations of ethical matters and questions, used extensively in this study, lies in the area of narrative. The practice of storytelling could well be a universal feature of nature-culture, including its function as a strong locus of ethical reflection and communication. The idea of narrative as a means of ethical deliberation is not new. As Martha Nussbaum has argued, there is a particular effectiveness of narrative as a means of ethical discourse, employing as it does the complexity of life as it is lived and the response of empathic imagination in hearers and readers.[5] As we saw in the last chapter, Katie Cannon has claimed narrative as a primary locus of ethical discourse for African American women. The contention here goes beyond the virtues of narrative as ethical discourse, however. The claim is that narrative can be a productive and just source for comparative ethics across cultural and religious difference. As Greg Sarris demonstrates, a wide range of stories, not just those with a recognizable structure of beginning, middle, and end, "can become a basis for *intercultural* and interpersonal communication and understanding" (emphasis added).[6]

One example of how narrative can work in very pragmatic ways toward this end is David Wellman's groundbreaking work, *Sustainable Diplomacy.* [7] In a case study of relations between Spain and Morocco, Wellman models an alternative methodology in the area of international relations, a methodology that recognizes the ecological situatedness of countries and the religions of peoples in international affairs. He employs both religious and personal narratives to find the reality "on the ground" and points of common conversation regarding norms and values. Another prime example is the work of Marilyn Legge in discerning Canadian feminist ethics in literature.[8]

Yet the use of narrative in cross-cultural and interreligious comparison is not a straightforward matter. Unlike Legge's and Wellman's use of narratives, which are told along with the solid background of description of the natural-cultural context in which they are alive, stories extracted from a context can undergo a radical shift in meaning (a self-evident idea within the conceptual framework of "moral habitat"). Above all, we need to remain aware that stories "can work to oppress or liberate, to confuse or enlighten. So much depends on who is telling the story and who is listening and the specific circumstances of the exchange."[9] Readings abstracted from context could easily replicate the colonialism and ethnocentrism of dominant practices of both philosophical and religious ethics in the West. Despite those risks, narrative is such a promising avenue for comparative ethics that it should not be dismissed or abandoned on that account. What is called for is an expansion of dialogue around how narratives are responsibly engaged and interpreted cross-culturally. That scrutiny and the further development of theory that would come out of it are well worth pursuing. What constitutes responsible practice in this area could use further definition. The pursuit of that definition ought to involve contributions from multiple social, religious, and geographic locations.

A caution similar to the observation that stories can oppress or liberate needs to be brought forward about all the practices that shape ethos. To make the observation that practices both express and shape ethos does not mean taking the "is" of an ethos for an "ought." No agent is as absolutely free as postulated by the Enlightenment, not even the European and Euro-American men who thought they were describing their own moral lives as rational agents making choices in total freedom. Yet the greater constriction of freedom in the lives of people of color, queer folk, and to a lesser extent white women and lower-class white men (not to mention otherkind), while something to be accounted for, is not to be placidly accepted as inevitable or justifiable by circumstance or "culture." Taking stock of the moral habitat as production of the whole biotic community of a location, and the "space" in which members of that community perform acts of agency, entails, for the privileged, clarifying the limitations under which we ought to exercise agency, and for all, contending with and contesting oppression. This brings me to the importance I want to claim for the task of reconceiving moral agency.

Moral Agency Reconsidered

Our present ethos might be more accurately described as an *im*moral, rather than a moral habitat. The entrenched structures of injustice, the momentum of destruction, the weaknesses of humankind—in the privileged global North our addictions to profligate and wasteful lives, our continual stoking of fears of the "other," and our seeming inability to face the truth and act politically, socially, and economically against one's own power and privilege—these are enough to drive the most optimistic among us into paralyzing despair. The moral agency that is now called for is one that is capable of subverting those powerful forces bent on destruction of Earth. Yet it seems well beyond our capacity to act to make any significant change, so much so that even the minimal political participation of voting in North America has been waning. And it is beyond our capacity, if we continue to think of ourselves as agents who act individually and independently, who can draw on nothing more than our individual character and moral fortitude.

In Cynthia Moe-Lobeda's discussion of where Christians can find the strength and courage to resist the forces of destruction of our lives and our planet, she identifies four dimensions of a subversive moral agency that recall the discussion of the previous chapter. This points to the need for such fulsome descriptions of human moral agency and the promise of this path of inquiry in cross-cultural and interreligious terms. Drawing from Luther's concept of the indwelling Christ, Moe-Lobeda starts with the claim that moral life with that centerpiece is rooted in "direct, sensuous, transforming relationship" (rather than cognitive belief), that this relationality includes all of creation, as well as God, that it is personal rather than abstract, and that it is incompatible with an apolitical and "private" spirituality.[10] I would suggest that pursuing further the question of moral agency of the *rest of creation* in relation to the Lutheran concept of indwelling Christ offers another way for Christian theology to conceive of Christ's presence in creation without pantheism, given that this description of moral agency insists on relationality among persons and not collapse of identity.

Conceiving moral agency in this fashion, as I do after my own study, is ultimately aimed at the development of embodied ethē, networks of meaning- and value-laden, value-forming practices that promote the survival, even flourishing, of the whole Earth community. All of this demands that ethics pay attention to ethos itself as a necessary focus of justice making and sustaining change agency, constantly examining the practices of a community, taking a role in altering destructive practices and cultivating the participation of all of its members. Personal ethics cannot be isolated as a separate area of concern from social ethics. Attention to individuals and acts cannot ignore their situatedness in moral habitats, in terms of opportunities and motivations, but also consequences. Nor can change be conceived as individual conversion. Change agency is not the cumulative actions of individuals but the exponential power of shifting expectations, meanings, and norms.

Just as oppression of other human beings points toward an unsustaining and unsustainable ethos, the absence and effacement of the contribution of otherkind to a community likewise distorts that community's character and the moral agency of human beings, as well as others, within it. The presence of otherkind means more than their proximity, it means their presence as agents who act for their own good, as well as others', who shape our moral imaginations, and who make claims on us. The actualization of this presence involves our attention. It is a matter of encounter, of relationship. That potential of relationship is thwarted, made exceedingly unlikely, in both the physical absence of otherkind or in the reduction of another's wholeness to our instruments, as it would be in the absence or reduction of human others to instruments for our "use." This is not only a defect of urban life, but also of rural life that instrumentalizes the living world.

As Robert Michael Pyle has observed, we cannot love or sustain that which we do not *know*. "What we know, we may choose to care for. What we fail to recognize, we certainly won't."[11] Both the physical absence of otherkind from our daily lives and instrumentalism block our knowledge of the Other and the Other's place—and ourselves and our places—by confusing purpose and meaning, and by defining both others and ourselves in terms of purpose, rather than being open to their self-generated disclosure of meaning. Moral habitat, to use Anna L. Peterson's words, "calls us to respect, take seriously, and seek out the viewpoints and the worlds shaped and inhabited not just by other humans but by a whole host of organisms sharing the planet. All these organisms are, like humans, embodied and embedded in the physical world. However, in various ways they are also all shapers of it, active agents and not merely blank paper waiting for human symbols and discourse (and hoes and bulldozers) to make something of them."[12]

Doing ethics in a moral habitat approaches territory traditionally covered under the rubric of character ethics, but from a different, more comprehensive direction. Ethos is the character of a *community*, the moral langue in which moral agency is parole. This larger scope of the moral life does not at all reduce the individual to an automaton or detract from the wholeness of a person as an organism, nor does it obviate the need for reflection and evaluation, but recognizes that individual moral agency is formed, enabled, and constrained by a larger whole. This places an individual's awareness, intent, and action always in the context of the patterns of power and justice in a community. And it suggests ways to think about *intergenerational and collective* agency and thus, inter-generational and collective responsibility. From here we can consider the collective nature of the moral life and moral formation in new ways, rec-ognizing both the importance of community and that the community is more-than-human.

The following discussions of intergenerational responsibility, abortion, and genetic engineering are not intended to be all-encompass-ing treatments. For each of them, this would entail a separate book. I am merely indicating by example how the ideas in this book can lead to new possibilities.

Directions in Intergenerational Responsibility

The discussion of human moral agency in the previous chapter drew from womanist reflection and indigenous traditions the idea that responsibility could be unearned. This sits uncomfortably with the dominant American culture. Because responsibility and culpability are so often confused, and because of the history of harm done to both African Americans (and African Canadians) and the indigenous peoples of the Americas, this is often heard as accusation, an unfair assignment of blame. This can be true when talking about the retrospective judgment of actions in the past that in the ethos of their time were done with the "best of intentions." It can also arise when talking about reparations or compensations for past harms, such as residential schools, land claims, lack of treaty recognition, or slavery. Unfortunately, the dominance of the cultures which originated in Europe means that such matters are often forced into their moral and legal frameworks. The courts of the dominant society are not the best place for redress. But unfortunately, the structures proposed or enacted as alternatives often operate unconsciously within the moral frameworks of the dominant society and remain subject to many unequally structured power arrangements.

One first step in trying to establish justice and right relationships in this process is to make the differences in moral habitats visible. An even more daring step would be to try to consider the matter through the frame of intergenerational responsibility. To indicate the promise of such an approach, I will relate a simple anecdote. In the wake of a decision by the Supreme Court of Canada that recognized a Mi'kmaq treaty right to commercial fisheries, violence broke out in several communities on the East Coast of Canada, most dramatically in the community of Esgenoôpetitj, or Burnt Church, New Brunswick. On the southwest coast of Nova Scotia, non-Native fishers were organizing to attack the traps and boats of Native fishers from the community of Bear River. On the night before a planned action, the chief of Bear River, the leader of the local Acadian fishers, and the provincial fishers' organization sat together in a room by themselves. The chief held an eagle feather and explained the custom of the talking circle, in which whoever held the feather had the floor. He asked that each of the three men present speak in two rounds. In the first round, they would try to speak what their ancestor seven generations removed would have to say, and in the second round, what their descendents seven generations hence would say. When the feather was handed to the Acadian, he said that his ancestor would say to the chief, "Your people kept us alive." In the second round, the three men considered the importance of preserving the viability of the fisheries for the future. Thinking intergenerationally and in the context of a more-than-human community offered a different place to consider a serious conflict which the courts had handled less than satisfactorily. Violence was averted. This story, which was later related on the CBC program *Ideas*, prompted a request from an Arab-Jewish kibbutz to have a visit from this chief.[13]

Directions in Abortion

First, a reminder that the intent here is not to resolve this issue or to give a definitive position, but to reframe the issue somewhat, in order to introduce more possibilities and authenticity into the discussion. Nor am I claiming that all the ideas here are original. I am simply illustrating how the ideas discussed in earlier chapters of this book open different ways to proceed than the usual debates. The discussion of abortion usually hinges on definitions of the point at which human personhood is established and the arbitration of the relative rights of a woman and a fetus. While these are important matters, exclusive attention to these questions isolates individuals and focuses entirely on the moment of decision about whether to continue a pregnancy or not. Some discussions will include the manner in which the pregnancy occurred, such as rape, or the likelihood of the occurrence of death due to illegal and unsafe abortions when access to abortion is not provided. But primarily the individual moment of decision is still the framework, and absolutist positions silence more nuanced considerations.

It is difficult to approach such a polarized issue. But a much different conversation arises when there is a recognition that a reduction in the need for abortions would be a desired good. It is perhaps only a partial good to those who oppose it absolutely, but it is a place to begin. A more fulsome discussion will acknowledge the symbolic value of the fetus for those who absolutely oppose abortion and engage a discussion on the level of moral imagination about the value of life in a more comprehensive sense than merely the time up until birth. The ethos of sexuality and the relative power of men and women in terms of setting the boundaries of sex and having access to social and economic supports and birth control are crucial points to examine. The social and economic reality beyond the individual woman or her immediate family needs to be kept in view. Finally, human population issues cannot be considered in isolation from the need for balance with the life needs of other species and the disparities of wealth globally. Responsibilities of a community to support parents and the responsibility of parents to consider the impacts of their choices to procreate on the future of the planet ought to be part of the discussion.

A growing number of voices are urging that the community to which the mother belongs is relevant to the question of abortion. Without taking away the ultimacy of the decision from a woman, the context in which she makes the decision needs to be understood. Can she be assured of adequate pre- and postnatal health care? Is there accessible child care and emotional support, if the mother needs to work? Or, is she married with children already and cannot reasonably support an additional child (a very common case)? Are families supported by more than ideological lip service—are workplaces accommodating for the needs of parents, is parental leave part of the social norm, are there opportunities to earn a living wage, so that one working parent can support a family, leaving another free to care for children, or is this the privilege only of an elite? What are the possibilities for nurturing education that will develop children's minds, hearts, and bodies? The well-being of families is a complex matter and part of the entire environment in which the question of abortion desperately needs to be set.

Directions in Genetic Engineering

In the previous chapter, I introduced a different set of categories from Nature and Culture, that of instrumental and noninstrumental. In the first category, instrumental, agency was located in groups of humans. But in the second, noninstrumental, as we have explored in this work, we can think about agency (even if limited and only provisionally moral in some cases) as inherently belonging to those in the second category in themselves. This can make a difference in many ways, including how we think about what we do with genes. The genetic engineering industry sometimes defends its work with transgenic experiments (those in which a gene of another species is introduced into an organism) as mimicry of "nature's" own "experiments" with transposed genes. One example given is that of the *Agrobacterium tumefaciens*, a common soil bacterium, which causes crown disease by transferring some of its DNA to the plant host. The transferred DNA is stably integrated into the plant genome, where its expression leads to the synthesis of plant hormones and thus to tumorous growth of crown gall.

Never mind that the transkingdom gene-swapping example given in the textbook is one that causes disease; it is assumed that human genetic intervention is one that will "improve" the "product." Distinguishing the instrument (which derives its meaning from purpose) from the noninstrumental (that which was created/has evolved independently of what it was "good for") and locating moral agency in that second category can be helpful in distinguishing different kinds of work with genes. The creation of new strains through cross-fertilization works, in a sense, *with* genes, collaborating with them in the things they generally do for their own good, not invading them. Genetic engineering, on the other hand, imposes on genes manipulative processes, forcing into them (raping them with?) elements they would not incorporate on their own. And where those genes express organisms with the ability to think, feel, and communicate, we need to engage, honor, and protect those capacities in the process of what we do with genes.

But this approach is insufficient in itself in doing ethics in a moral habitat. We also need to set our actions in the context of the wholes beyond the organism. Here is where the second distinction mentioned above— internal to the instrumental category but referencing the noninstrumental— becomes crucial. In making this distinction I follow Aidan Davison's argument that we need to work very hard at determining which technological practices are not only *sustainable* in terms of ecological efficiency, but which are *sustaining* of noninstrumental life, including our own.[14]

Inserting bacterial genes, such as BT, into corn and potatoes not only violates the integrity of corn and potatoes, it can disrupt the life processes of insects beyond those that were the reason for the genetic manipulation in the first place. Moreover, the indiscriminate spread of these bacterial organisms can cause the development of immunity in the target insects, since that is what their genes normally do, and thus deprive organic growers of a nonchemical ally.

Though we need to draw our sustenance from the members of the category of beings we have not generated, we need to take care not to assume that we have been authorized to move them arbitrarily and completely from one category to another or thrust them into the borderlands for whatever purpose we choose. We need to remember that we ourselves are inside this noninstrument category, not standing outside. We did not create ourselves. The digerati have already raised the question that comes in the wake of blithely considering all as means and at our disposal. Given our conceivable capabilities to create machines that equal or surpass us in accomplishing nearly every cognitive and mechanical task, that can learn and move and reproduce themselves, Kelly says the central question of the twenty-first century will be "What are humans good for?"[15] Good *for*?

Even the constructions of instrumental/noninstrumental boundaries, however, are insufficient by themselves; just because something is noninstrumental in its origin does not put a check on our making it so, through ideas or our manipulation of it. We need the witness of a cosmology (like that of the Priestly tridents) in which the generative power of the universe declares all heavenly and earthly bodies simply "good," not "good for," as an important source of wisdom from outside this world of self-referential technoscience. We need the witness of a universe filled with persons who exchange gifts of life and are bound to each other and to all in ties of obligation as an important source of wisdom.

Doing ethics in this technoscience habitat means breaking the stranglehold of its patterns of self-reference. Science demonstrates in its own way that we are not occupants of Earth but as Thomas Berry puts it "dimensions of the earth and indeed the universe itself." It is to that larger whole, and in particular to dimensions of it that science may not be able to grasp, that we need to turn with humility rather than arrogance. "We are an immediate concern of every other being in the universe. Ultimately our guidance on any significant issue must emerge from this comprehensive source."[16] Berry places before us two fundamental courses for humankind and the rest of Earth, now that we have acquired the capacities to alter life systems and thus, both with and without the consent of various peoples, intertwined the lives of all beings of Earth. We can move into what Berry terms a "technozoic" age or an "ecozoic" one. Berry does not elaborate very much on why we should avoid the technozoic, one in which our "sustainability" is secured by technological advances. He expects his readers, probably as in love with Earth's beauty and wildness as he is, to be equally skeptical of the promises that the technoscience practices that have been so destructive can somehow be the means of their own salvation. He is not opposed to technologies, but insists they need to be ordered to be coherent with the ever-renewing technologies of Earth itself.[17]

But as Haraway has shown in her example of the genetic engineering textbook and its portrayal of "nature" as "genetic engineer" and as the digerati demonstrate with their talk of "evolution" to a new substrata of "silicon-based life" and "spiritual machines," seeing Earth's dynamic interplay of energy/matter as self-renewing "technologies" can mean very different things depending upon where one is located. And so we need a distinction between the instrumental and noninstrumental, and we need clarity in the discernment of what is sustaining of life. We need the numinous presence of otherkind to help us. And we need the guidance of other sources of wisdom.

Toward a Future Worth Living In

Rather than a series of technical problems of material and social engineering that need technical solutions, the moment we are in now calls for the creativity of the universe itself, and for each of us who are dimensions of that universe to contribute with our whole selves, in the same spirit of collaboration, interdependency, and love of beauty and diversity that has brought us into being. Thomas Berry compares this task to that of an artist, who in creating a significant work "first experiences something akin to dream awareness that becomes clarified in the creative process itself."[18] For this, he says, we need the same kind of sustaining vision that an artist has in that dream-awareness. He proposes that vision as the "Ecozoic Era, the period when humans would become a mutually beneficial presence on the Earth."[19]

In many ways, we do not have the luxury of time that created the universe. The changes of climate instability bear down on humans and especially on plants with ferocious speed that may outstrip our vast adaptive capacities. This is nevertheless not only an urgent work but an immense and a long-term one. We need stamina as well as creativity. Berry calls it "the Great Work" of our time, "the task of moving modern industrial civilization from its present devastating influence on the Earth to a more benign mode of presence."[20] This is not a task we have chosen, but for which we have been chosen, according to Berry, in a way that resonates well with the idea of "unchosen responsibility" we found in womanist writings and among the Mi'kmaq. The vision of a community of relations that is inclusive of the more than human enables us to see ourselves, not as managers, but allies, in the healing of Earth with the creative power of the universe embodied in it. Such an immense task can have no single answer to the question of what we shall be and do, in terms of *how* we are to act in order to bring this about.

So I offer no grand plan, no manifesto. No metanarrative, no end of the story. But I do insist on this: Our aliveness to what Berry calls the numinous powers of the universe, to the presence of a wide variety of modes of personhood, to its capacity for sustaining us though exchanges among persons of gifts and obligations, call especially for attention to otherkind and to those whose understanding of moral agency is broad and deep. Intimacy with the rest of the Earth community is not so different from intimacy between human beings, or with God. It requires our vulnerability; and because, as Augustine remarked, a picture of food does not nourish, we need to be mutually present to the other with our bodies, minds, and hearts. To be intimate with the more-than-human world is not a matter of mere proximity, but demands the "work of attention" Murdoch speaks of, and this is something our habits of technologically representing the otherkind actually impair. I am concerned that as we become accustomed to "virtual" realities and the screen-saver version of "nature," we cut ourselves off from any context for otherkind "in ideas larger than 'fresh' or fierce or cute. And so we don't know what to make of them . . . hence most of our responses are artificial."[21] Surfeited with screen images, we lose patience with the live wild that is not interested in amusing us. That stings and nibbles at us. That puzzles, frightens, and bores.

Intimacy requires as well the respect for the autonomy of the other, while expressing a love that acts for the good of the other. It involves sacrifice, gratitude, and obligation. It calls for celebration and humor and grief.

Recognizing that we struggle not alone as individuals or communities of peoples, but as a Community of Life is one thing, learning *how* to seek the guidance and support of Others, human and more-than-human, on a day-to-day basis is quite another. One lesson of ethos as moral habitat is that while the scope of change that needs to be made is large, it is woven of all the strands of practices that produce, reproduce, and are produced as networks of norms, values, and meanings. Ethos as moral habitat says that reweaving happens exactly where we are, in whatever moral habitat we act as moral agents. It says that small movements reverberate through a great web. Those of us in industrial, technoscience habitats need to open ourselves in our own specific locations to the presence, guidance, and assistance of others whom we have exploited and excluded, and in this we may employ a vast variety of practices, but especially those that cultivate our own humility. In our own traditions, those practices we associate with the "spiritual," understood in a nondualistic way rooted in "earth and its distress," make space in our souls for those numinous powers to be present to us. They serve to clarify what we are to be and to do. With a task this immense, no act or gesture is too small to be unnecessary, nor large enough, by itself, to be sufficient.

Being satisfied too easily with insignificant change, the privatization of practices of spirituality, and the hubris that we have to do something grand or nothing, keep us isolated and impotent in unraveling and reweaving the sticky threads that constitute us as moral agents, that inhibit us as we attempt to stop the destruction and move toward sustaining ways of life. In the early 1980s Margaret Thatcher made the acronym TINA—"there is no alternative" (to a particular geopolitical, economic agenda)—famous, or infamous, depending on your point of view. Resistance is futile. And she was right as long as the alternative was conceived in the singular. While not as catchy as TINA, I would propose we think in terms of TAMA—there are many alternatives, alternatives that are both possible and necessary. What we need most now is not grand solutions proposed for everyone everywhere, as communism envisioned itself an alternative to capitalism. What we need now are the thousands of small steps that enable us to move with and as Earth against what is destroying Earth, to reflect Earth's own diversity and particularity of place.

To focus on the practices that weave an ethos, to take it step by step and consider our progress in terms of working with Earth where Earth wishes to go, might seem trivial and futile as a path of resistance. Yet as Davison says, "To reclaim even in small measure our practices from the promises of technological efficiency is to reclaim political possibilities for forms of practical decision making capable of honoring that which sustains us. It is to lift policy off its precariously narrow pedestal of instrumentalist and found it on a more fully human footing."[22] To that I would add a more fully more-than-human footing as well.

What I can tell you from the moral habitat of a permaculture garden is this. Yes, there are mighty and powerful forces behind the militaristic/capitalistic commerce of consumption, but the extensive weapons and the relentless onslaught of advertising and cultural extreme measures being brought to bear—promotion of destructive desires and practices—themselves testify to the groundlessness of their agendas. Were these agendas true and sustaining, they would not need such force to keep them in place. I am always amazed at the humble power of the plants in my garden. It is not the perfect place for them; conditions are rarely "just right." But they only need half a chance, so great is their thrust toward the sun, toward the flower, toward the fruit. However great the powers of destruction, they are no match for that which activates the seed. They are no match for the One who declares, "Let the earth sprout sproutlings . . ."

Notes

Introduction

1. George Santayana, "The Genteel Tradition in American Philosophy," in *The Genteel Tradition: Nine Essays*, ed. Douglas L. Wilson (Cambridge, MA: Harvard University Press, 1911/1967), 63.

2. 230d.

3. *The Genteel Tradition*, 63.

4. If culture consists of that information that is passed on "by behavioral rather than genetic means," then other animals also have culture. John Tyler Bonner, *The Evolution of Culture in Animals* (Princeton: Princeton University Press, 1980), 4.

5. The term "natural-cultural" or "nature-culture" (hyphenated) is taken from Bruno Latour, *We Have Never Been Modern* (Hertfordshire UK: Harvester Wheatsheaf, 1993). Larry Rasmussen similarly uses "nature/culture" to bridge categories. Larry Rasmussen, *Earth Community Earth Ethics* (Maryknoll, NY: Orbis, 1996), 46.

6. Max L. Stackhouse, *Ethics and the Urban Ethos: An Essay in Social Theory and Theological Reconstruction* (Boston: Beacon, 1972), 5.

7. *The Ethos of the Cosmos: The Genesis of Moral Imagination in the Bible*. (Grand Rapids, MI: Eerdmans, 1999).

8. Donald A. Grinde and Bruce E. Johansen, Ecocide of Native America: Environmental Destruction of Indian Lands and Peoples (Santa Fe: Clear Light, 1995); Maria Mies and Vandana Shiva, eds., *Ecofeminism* (Halifax, NS, London, and New Jersey: Fernwood/Zed, 1993).

9. A prime example of this is Shepard Krech III, *The Ecological Indian: Myth and History* (New York: Norton, 1999). While Krech points

111

out that embedded in the concepts of "ecological" and "environmental" are "certain cultural premises about the meanings of humanity, nature, animate, inanimate, system, balance, and harmony, and their suitability for indigenous American Indian thought or behavior should not be taken as a given" (22), he nevertheless proceeds to evaluate findings of Euro-American archeological and historical scholarship about indigenous Americans on the basis of these eurocentric concepts.

10. David Abram, *The Spell of the Sensuous: Perception and Language in a More-Than-Human World* (New York: Vintage/Random House, 1997).

11. Charlene Spretnak, *The Resurgence of the Real* (New York: Routledge, 1999), 3.

12. In Euro-American terms, I immigrated to Canada, to the province of Nova Scotia. Mi'kmaki covers that Euro-American political territory, plus the province of Prince Edward Island, and parts of New Brunswick, Newfoundland, Quebec, and the state of Maine.

13. I have decided to follow the convention for Mi'kmaw/Mi'kmaq described by linguists Bernie Francis and Virick C. Francis, using Mi'k-maw in the singular and Mi'kmaq in the plural. Where styles used by others differ, I retain the original; therefore Mi'kmaw, Mi'kmaq, and Micmac should be understood as having the same meaning. The Mi'k-maq are an indigenous people of Atlantic Canada.

14. Carol A. Newsom, "Common Ground: An Ecological Reading of Genesis 2–3," in *The Earth Story in Genesis*, ed. Norman Habel and Shirley Wurst (Sheffield, UK and Cleveland, OH: Sheffield/Pilgrim, 2000), 60.

15. Mary Midgley, *Science as Salvation: A Modern Myth and Its Meaning* (London and New York: Routledge, 1992), 1.

16. Actually, Descartes was not as extreme in his application of the mechanistic metaphor as was Hobbes, for whom the entire universe was mechanical, including humans. Descartes preceded him, however, and was particularly influential.

17. Donna J. Haraway, "Situated Knowledges: The Science Question in Feminism and the Privilege of Partial Perspective," in *Simians, Cyborgs, and Women: The Reinvention of Nature* (New York: Routledge, 1991).

18. *Europe and the People without History* (Berkeley: University of California Press, 1982), 3.

19. Yi-Fu Tuan, *Topophilia: A Study of Environmental Perception, Attitudes, and Values* (New York: Columbia University Press, 1990).

Chapter 1. Ethos as Moral Habitat

1. Stackhouse, *Ethics and the Urban Ethos: An Essay in Social Theory and Theological Reconstruction.*

2. Aristotle, *Nicomachean Ethics*, 1139a1; Homer, *Iliad* 6.511, *Odyssey* 14.411.

3. Herodotus, 7.125; Oppianus, *Haleutica* 1.93.

4. Paul L. Lehmann, *Ethics in a Christian Context* (New York and Evanston: Harper & Row, 1963), 24. Personally, I don't find it humiliating, but rather refreshingly humbling, both of those words being etymologically rooted in the same ground, *humus.*

5. Hesiod, *Works and Days*, "abodes of humans" 167, 525; manners, customs, 137; character, 67, 78.

6. Herotodus, human homes, 1.15, 157, customs, 2.30, 35, 4.106.

7. *Historia Animalum*, 487a, 12.

8. Brown, *Ethos*, 11.

9. Ibid.

10. Charles Taylor, *Sources of the Self: The Making of the Modern Identity* (Cambridge, MA: Harvard University Press, 1989), 7.

11. Later I will deal with language itself as constitutive of moral habitat, but here I am merely building on an analogy. I will also argue for a more embodied and fluid understanding of culture/ethos than would be implied by this analogy.

12. Ferdinand de Saussure, *Cours De Linguistique Générale* (Paris: Payot, 1978).

13. "The Land Ethic," *A Sand County Almanac* (New York and Oxford: Oxford University Press, 1987/Orig. 1949), 204.

14. Ibid.

15. Rasmussen, *Earth Community*, 48.

16. Brown, *Ethos*, 10.

17. Ibid., 28.

18. Spencer, *Gay and Gaia* (Cleveland, OH: Pilgrim, 1996), 300.

19. Ibid., 296 (emphasis original).

20. Abram, *The Spell of the Sensuous*.

21. Clifford Geertz, *The Interpretation of Cultures: Selected Essays* (New York: Basic, 1973), 89. Geertz notes in brackets that why this should be more problematic with culture than the similar situations of "social structure" or "personality" is "something I do not entirely understand."

22. Ibid.

23. Susanne Katherina Knauth Langer, *Philosophical Sketches* (Baltimore: Johns Hopkins University Press, 1962), quoted by Geertz, *The Interpretation of Cultures*, 89.

24. Geertz quite explicitly identifies culture and social structure as "abstractions from the same phenomena," *Interpretation of Cultures,* 145.

25. This kind of disembodied definition of culture as a pattern of meaning or system of symbols or concepts has ramifications. To think similarly of religion as a system of belief is typical in modern Western thought. It makes possible the kind of thinking in the decision in the U.S. Supreme Court case of *Lyng v. Northwest Indian Cemetary Protection Association.* In writing for the majority, Justice Sandra Day O'Connor made such a distinction between belief and practice of religion, claiming that the Constitution's protection of freedom of religion applies only to belief, not the practice of a religion. Therefore it was not a violation of four First Nations' freedom of religion to allow a road to be constructed through their burial grounds and sacred lands.

26. Clifford Geertz, *Local Knowledge: Further Essays in Interpretive Anthropology* (New York: Basic, 1983), 3.

27. *Interpretation of Cultures,* 17.

28. Ibid., 112.

29. Ibid., 127.

30. Ibid.

31. See Mary Midgley's discussion "Facts and Values," in Mary Midgley, *Beast and Man* (Ithaca, NY: Cornell University Press, 1978), 177–200.

32. *Interpretation of Cultures,* 67–68.

33. *Ethos of the Cosmos,* 10. Brown draws this expression of culture as craftwork from S. Hart, "The Cultural Dimension of Social

Movements: A Theoretical Reassessment and Literature Review," *Sociology of Religion* 57 (1996), 98.

34. Vandana Shiva, *Monocultures of the Mind: Perspectives on Biodiversity and Biotechnology* (London, UK; Atlantic Highlands, NJ; Penang, Malaysia: Zed/Third World Network, 1993).

35. Akhil Gupta and James Ferguson, "Culture, Power, Place: Ethnography at the End of an Era," in *Culture Power Place: Explorations in Critical Anthropology*, ed. Akhil Gupta and James Ferguson (Durham and London: Duke University Press, 1997), 5.

36. *Ethics and the Urban Ethos*, 5.

37. *The Interpretation of Cultures*, 127.

38. Stephanie Lahar, "Roots: Rejoining Natural and Social History," in *Ecofeminism: Women, Animals, Nature*, ed. Greta Gaard (Philadelphia: Temple University Press, 1993), 96. As quoted by Daniel T. Spencer, *Gay and Gaia*, 296–97 (emphasis Spencer's).

39. Richard Dawkins, *The Selfish Gene* (Oxford: Oxford University Press, 1976). Dawkins's later works employ the concept of "meme" as an analogous entity to the gene at the social/cultural level, in a way that might be thought similar to the role I am conceiving for ethos. The degree of negation of agency at the level of the individual, however, as well as the problematic persistence of the metaphor of the machine, are critical differences.

40. Lawrence E. Johnson, *A Morally Deep World: An Essay on Moral Significance and Environmental Ethics* (Cambridge/New York: Cambridge University Press, 1991).

41. Dawkins, *Selfish Gene*, x.

42. Edward O. Wilson, *On Human Nature* (Cambridge, MA: Harvard University Press, 1975), 167.

43. Midgley, *The Ethical Primate*.

44. Public lecture, November 16, 2000, American Museum of Natural History, New York, New York. Eldredge said that he has come away from these conversations with the impression that the age of the earth is not an obstacle, although it would conflict with a literalist reading of the scripture. The real difficulty lay in the kinship with "lower" animals, for this would remove any source of morality. While I have to work hard to understand what the difficulty is, he thought it tremendously significant.

45. Midgley, *Beast and Man* (New York: Cornell University Press 1979), xix.

46. *The Ethical Primate*, 115.

47. Paul Shepard, *The Others: How Animals Made Us Human* (Washington, DC: Island Press, 1997).

48. Ibid., 18.

49. Ibid., 20.

50. Ibid., 20–21.

51. Ibid., 22 (emphasis added).

52. Ibid.

53. Lakoff and Johnson, *Philosophy in the Flesh: The Embodied Mind and Its Challenge to Western Thought* (New York: Basic, 1999), 17.

54. Shepard, *The Others*, 46.

55. For a discussion of variable meanings of mountains, and how changes in meaning are related to environmental attitudes, see Tuan, *Topophilia: A Study of Environmental Perception, Attitudes, and Values*, 70–74.

56. Shepard, *The Others*, 24.

57. Lakoff and Johnson, *Philosophy in the Flesh* (New York: Basic, 1999); Mark Johnson, *Moral Imagination: Implications of Cognitive Science for Ethics* (Chicago: University of Chicago Press, 1993).

58. *Philosophy in the Flesh*, 43–44.

59. Ibid., 323.

60. Ibid., 290–301.

61. *Moral Imagination*, 33.

62. *Philosophy in the Flesh*, 290–91.

63. *Beast and Man*, 182.

64. Ibid., 182–83.

65. Ibid.

66. *Traces on the Rhodian Shore*, 87.

67. Hippocrates, *Airs, Waters, Places,* Translated from Greek by W. H. S. Jones, Loeb Classical Library (Cambridge, MA: Harvard University Press, 1948 [1923]), vol. 1 of *Works of Hippocrates*, xxiv. Quoted by Glacken, *Traces on the Rhodian Shore*, 87. Glacken notes that this was one of several leading ideas in this work of Hippocrates, but while his themes included cultural influences such as institutions on the char-

acter of a folk, this essay's influence was overwhelmingly due to the theme of the close relationship between culture and environment.

68. E.g., Griffith Taylor, *Environment and Race* (Oxford: Oxford University Press, 1927).

69. Lucinda A. McDade et al., eds., *La Selva: Ecology and Natural History of a Neotropical Rainforest* (Chicago: University of Chicago, 1994).

70. *Topophilia*, 80–81.

71. "Landscape, History, and the Pueblo Imagination," in David Landis Barnhill, ed., *At Home on the Earth: Becoming Native to Our Place: A Multicultural Anthology* (Berkeley and Los Angeles: University of California Press, 1999), 39.

72. Ibid., 38.

73. Ibid., 40.

74. Ibid., 42.

75. Iris Murdoch, *The Sovereignty of Good* (New York: Schocken, 1971), 37.

76. Thomas E. McCollough, *The Moral Imagination and Public Life: Raising the Ethical Question* (Chatham, NJ: Chatham House, 1991), 16–17.

77. Rasmussen, *Earth Community*, 32

78. Ibid., 7.

79. Adrienne Rich, *What Is Found There: Notebooks on Poetry and Politics* (New York: Norton, 1993).

80. Haraway, "Situated Knowledges: The Science Question in Feminism and the Privilege of Partial Perspective,"198.

Chapter 2. "The Great Community of Persons"

1. Cronon, William. *Changes in the Land: Indians, Colonists, and the Ecology of New England* (New York: Hill & Wang, 1983), 13.

2. Richard J. Preston, "The Great Community of Persons," in *Papers of the Twenty-Eighth Algonquian Conference*, ed. David H. Pentland (Winnipeg: University of Manitoba, 1997).

3. James B. Waldram, "The Reification of Aboriginal Culture in Canadian Prison Spirituality Programs," in *Present Is Past: Some Uses of Tradition in Native Societies*, ed. Marie Mauze (Lanham: University Press of America, 1997), 131–43.

4. Marie Battiste, "Cultural Transmission and Survival in Contemporary Micmac Society," *The Indian Historian* 10, no. 4 (1977): 13.

5. Lisa Philips Valentine, "Performing Native Identities," in *Actes Du Vingt-Cinquième Congrès Des Algonquinistes*, ed. William Cowan (Ottawa: Carleton University, 1994), 483.

6. Murdena Marshall, "Salite: A Mi'kmaq Sacred Tradition," in *The Mi'kmaq Anthology*, ed. Rita Joe and Lesley Choyce (Lawrencetown Beach, NS: Pottersfield, 1997).

7. Although this linguistic taxonomy is Euro-American in origin, it is used and built upon for cultural and political alliances among the First Nations peoples themselves, so I do not hesitate to use it.

8. Cronon, *Changes in the Land*, p. 62. For an in-depth look at the complex formative factors for Northern Algonquian peoples, see A. Theodore Steegmann, Jr., ed., *Boreal Forest Adaptations: The Northern Algonkians* (New York: Plenum, 1983).

9. Early contact Mi'kmaq society was not totally egalitarian; nevertheless there are striking differences in social stratification with Northwest Coast peoples, whose habitat, though also maritime, is different in critical ways. A useful comparison demonstrating how differences in coastal habitat are reflected in social structure has been done by R. G. Matson, "Intensification and the Development of Cultural Complexity: The Northwest versus The Northeast Coast," in Ronald J. Nash, ed., *The Evolution of Maritime Cultures on the Northeast and the Northwest Coasts of America* (Vancouver: Simon Fraser University, 1983). On the traditional sociopolitical organization of Mi'kmaq society see in that same volume Virginia P. Miller, "Social and Political Complexity on the East Coast: the Micmac Case," 41–55, and Daniel N. Paul, *We Were Not the Savages: A Micmac Perspective on the Collision of European and Aboriginal Civilizations* (Halifax, NS: Nimbus, 1993).

10. See Howard L. Harrod, *The Animals Came Dancing: Native American Ecology and Animal Kinship* (Tucson: University of Arizona Press, 2000).

11. Marie Battiste, "Nikanikinu'tmaqn." In *The Mi'kmaw Concordat*, ed. James Sa'kej Youngblood Henderson (Halifax: Fernwood, 1997), 15. Battiste may be using "person" here to refer to human people alone when she says, "Just as a person has a life-force, so does a plant, rock or animal." This is likely an accommodation to her readers' English, since the spark of life is shared, and one part of a tripartite "soul" characteristic of all beings, *wjijaqami*, includes the prefix indicating personal, or personhood.

12. Hunting was a largely gender-specific practice.

13. Battiste, "Nikanikinu'tmaqn," 15.

14. Harvey A. Feit, "The Ethno-Ecology of the Waswanipi Cree; or How Hunters Can Manage Their Resources," in *Cultural Ecology*, ed. Bruce Cox (Toronto: Macmillan, 1978), 115–25. Feit's study covered a period in the late 1960s, so this framework is by no means to be considered superceded by a modern Western one, at least among those nations who have been able to maintain a hunting/foraging primacy in their sustenance. An outstanding treatment of Algonquian hunting ethos is Adrian Tanner, *Bringing Home Animals: Religious Ideology and Mode of Production of the Mistassini Cree Hunters* (St. John's: Memorial University of Newfoundland, 1979). Some anthropologists, including Feit, report that the agent of the gift was an "animal master," a spiritual being who looked after the welfare of a certain species, or that the wind was the giver. It may be that all three—individual animal, "master," and wind—could be considered possible benefactors at any particular time. And while personhood is shared across species, there is a strong taboo against cannibalism in the stories of the Windigo.

15. Orechia, *Maliseet and Micmac*, 52.

16. A. Irving Hollowell, "Ojibwa Culture and World View," in *Contributions to Anthropology: Selected Papers of A. Irving Hallowell* (Chicago and London: University of Chicago Press, 1976), 363.

17. Ibid.

18. Battiste, "Nikanikinu'tmaqn," 15.

19. Shape-shifting is a prominent feature of Algonquian worldviews. Many animals in Mi'kmaq and other Algonquian stories take the form of human beings when interacting with them, and humans shapeshift also. Some anthropologists who have studied Algonquian cultures think shape-shifting applies only to certain kinds of humans and to animal spirit-people, rather than the ordinary individual animal itself, but the narratives of the Mi'kmaq make no such distinction.

20. As recorded by William H. Mechling in the early twentieth century, *Malecite Tales* (Ottawa: Government Printing Bureau, 1914), cited in Orechia, *Maliseet and Micmac*, 103.

21. Basil Johnston, *Ojibway Heritage* (Toronto: McClelland & Steward, 1984).

22. Bernard G. Hoffman, *The Historical Ethnography of the Micmac of the Sixteenth and Seventeenth Centuries*. PhD Thesis. University of California, 1955.

23. The differences might reflect different experiences in acquiring corn as it disseminated from the south, the stories of other peoples who

acquired corn earlier, or the relative difficulty in growing conditions for corn of various areas (some areas in Algonquian regions have very short growing seasons, in which corn would be marginal).

24. An excellent contemporary description of such a process is given by Douglas Cardinal, "Dancing with Chaos: An interview with Douglas Cardinal," in *Intervox*, vol. 8 (1989/90): 27–31, 44–47, reprinted in McPherson and Rabb, *Indian from the Inside*, 67–81.

25. Ruth Holmes Whitehead, *Stories from the Six Worlds: Micmac Legends* (Halifax, NS: Nimbus, 1988).

26. Sunset Rose Morris, "Spring Celebration," in *The Mi'kmaq Anthology*, ed. Rita Joe and Lesley Choyce (Lawrencetown Beach, NS: Pottersfield, 1997).

27. Marie Battiste, "Mi'kmaq Socialization Patterns," in *The Mi'k-maq Anthology*, ed. Rita Joe and Lesley Choyce (Lawrencetown Beach, NS: Pottersfield, 1997), 148. Also Murdena Marshall, "Values, Customs and Traditions of the Mi'kmaq Nation," *The Mi'kmaq Anthology*, pp. 51–63.

28. Battiste, "Nikanikinu'tmaqn," 15.

29. Calvin Martin, *Keepers of the Game: Indian-Animal Relationships and the Fur Trade* (Berkeley: University of California Press, 1978), 186. Martin later demonstrates an approach different from that of this book. Although not one that would contradict the particular point being made here, it is worth reading his later book for balance: Calvin Luther Martin, *The Way of the Human Being* (New Haven and London: Yale University Press, 1999).

30. J. Baird Callicott, *In Defense of the Land Ethic: Essays in Environmental Philosophy* (Albany: State University of New York Press, 1989), 216.

31. This became sharply evident in the rhetoric surrounding the resumption of ceremonial whaling by the Makah in 1998.

32. *The Others*, 36.

Chapter 3. Agents of and Respondents to God

1. Gerhard von Rad, "The Theological Problem of the Doctrine of Creation," in *Creation in the Old Testament*, ed. Bernhard W. Anderson, Issues in Religion and Theology (Philadelphia and London: Fortress/SPCK, 1966), 56. Von Rad's whole essay is more subtle, and

at a later point he moved away from his earlier positions. My point here is to draw attention to the fact that the idea of creation and history as two different subjects has been imposed, and mightily influential on how we as late moderns read, not that it is the only way these texts have been read.

2. Ibid. Von Rad is honest enough to observe how, in Deutero-Isaiah "We are struck by the ease with which two doctrines, which to our way of thinking are here brought together" (p. 57). He does not, however, allow this to interrupt for very long his thesis that creation is separate and subordinate.

3. Theodore Hiebert, *The Yahwist's Landscape: Nature and Religion in Early Israel* (New York/Oxford: Oxford University Press, 1996), 5.

4. Ibid.

5. Hiebert, *The Yahwist's Landscape*, 11–12.

6. This framework may not be logically consistent with itself (never mind the correspondence to the actual conditions of ancient Israel), but that did not hamper its explanatory power for moderns, within their worldview.

7. Hiebert, *The Yahwist's Landscape*, 12.

8. Even in his argument that the detachment of creation and history as separate doctrines "violates the intention of the creation stories" (p. 33), Bernhard W. Anderson accepts history as primary and subsuming of the category of creation, and promotes the scholarly consensus that "It may now be said that it is the biblical sense of history which accounts for the singularity of Israel's faith in relation to other religions, ancient or modern. Other peoples of antiquity, to be sure, had some awareness of the dimension of history and could even speak of their gods as taking part in historical events. But by and large the religions of Israel's neighbors were tied to the sphere of nature, where the cyclical rhythms were determinative for man's (*sic*) existence. Israel parted with the religions of the ancient Near East by declaring that history is the area of ultimate meaning precisely because God has chosen to make himself known in historical events and to call men to participate in his historical purpose. If today we share, to some degree, this historical consciousness—even in secularized versions—we are primarily debtors to the Israelites and their Christian heirs, not to the Babylonians, Egyptians, Canaanites or Greeks" (p. 27). Bernhard W. Anderson, *Creation versus Chaos: The Reinterpretation of Mythical Symbolism in the Bible* (New York: Association, 1967).

9. G. Ernest Wright, *God Who Acts: Biblical Theology as Recital* (London: SCM, 1952). Those scholars contesting that there is a dichotomy of creation and history include Terence E. Fretheim, "The Reclamation of Creation: Redemption and Law in Exodus," *Interpretation* 45 (1991): 354–65; and the previously mentioned work by Brown, *The Ethos of the Cosmos* and Hiebert, *The Yahwist's Landscape*, 3–29.

10. Thomas Berry and Thomas Clarke, *Befriending the Earth: A Theology of Reconciliation between Humans and the Earth* (Mystic, CT: Twenty-third Publications, 1991), 75.

11. Gene M. Tucker, "Rain on a Land Where No One Lives: The Hebrew Bible on the Environment," *Journal of Biblical Literature* 116, no. 1 (1997). See n. 4, p. 4 for a representative list of the literature to that date.

12. *Ethos*, 5.

13. Not all scholars writing out of an ecological concern recognize the extent to which the nature/culture dichotomy still frames their interpretation. As William P. Brown has pointed out, many of these writers assume that "creation" consists of nonhuman natural elements exclusively, and the concern to "heal creation" lacks a "concomitant concern for healing the community." *Ethos*, 3. So while some ecologically oriented writers are questioning the centrality of human history as a given, scripturally, not many are able to suspend the creation/history dichotomy as an appropriate heuristic for the Bible.

14. Martin LaBar, "A Biblical Perspective on Nonhuman Organisms: Values, Moral Considerability, and Moral Agency," in *Religion and the Environmental Crisis*, ed. E. Hargrove (Athens: University of Georgia Press, 1986), 76–93.

15. "Self-Consciousness and the Rights of Nonhuman Animals and Nature," *Environmental Ethics* I (1979), 100.

16. LaBar, "A Biblical Perspective," 88.

17. Ibid., 89.

18. Ibid.

19. Average annual rainfall rates vary from four inches (100 mm) in the Dead Sea area to twenty-eight inches (700 mm) in Upper Galilee. And the ideal timing and amount of rainfall for crop production occurs about a third of the time. "Geography and Ecology of the Land of Israel," in John Rogerson and Philip Davies, *The Old Testament World* (Englewood Cliffs, NJ: Prentice Hall, 1989).

20. S. Dean McBride, Jr. "Divine Protocol: Genesis 1:1–2:3 as Prologue to the Pentateuch," in *God Who Creates: Essays in Honor of W. Sibley Towner*, William P. Brown and S. Dean McBride, Jr., eds. (Grand Rapids and Cambridge: Eerdmans: 2000), 3–41.

21. Ibid.

22. Norman Habel, "Geophany: The Earth Story in Genesis," in *The Earth Story in Genesis*, ed. Norman Habel and Shirley Wurst, Earth Bible (Sheffield, UK and Cleveland, OH: Sheffield/Pilgrim, 2000). Habel identifies Earth as present, but hidden by the darkness and waters, waiting to be revealed, hence, Genesis 1 as geophany.

23. Nahum M. Sarna, *Genesis: The Traditional Hebrew Text with New Jps Translation/Commentary, The Jps Torah Commentary* (Philadelphia, New York, Jerusalem: Jewish Publication Society, 1989).

24. *Ethos*, 39. Brown's earlier study of the earliest textual tradition, as reflected in the old Greek, concluded that the waters had an even more active, generative role (v. 20) than in the later MT, with the upper waters generating winged life as the lower do sea life. William P. Brown, *Structure, Role, and Ideology in the Hebrew and Greek Texts of Genesis 1:1–2:3*, ed. David L. Petersen, vol. 132, *Sbl Dissertation Series* (Atlanta, GA: Scholars, 1993). See also William P. Brown, "Divine Act and the Art of Persuasion in Genesis 1," in *History and Interpretation: Essays in Honour of John H. Hayes*, ed. M. P. Graham et al. (Sheffield: Journal for the Study of Old Testament, 1993).

25. Mark C. Brett, "Earthing the Human in Genesis 1–3," in *The Earth Story in Genesis*, ed. Norman C. Habel and Shirley Wurst, *The Earth Bible* (Sheffield, UK and Cleveland, OH: Sheffield/Pilgrim, 2000); Terence E. Fretheim, "Creator, Creature and Co-Creation in Genesis 1–2," *Word and World Suppl* 1 (1992). Also Norman Habel, "Geophany"; McBride, "Divine Protocol," and discussions with Alan Cooper.

26. Jewish Publication Society, King James Version, New Revised Standard Version, New International Version all translate passively; cf. New Jerusalem Bible "Let the waters under heaven come together."

27. Niphal forms can also be used for resultative meaning, but that indicates a static state—where the verb "to be" is not an auxiliary as it is in the passive form, but the main verb, and "gathered" is an adjective—an unlikely meaning here.

28. Brown, *Ethos*, 40.

29. Ibid.

30. Ibid., 41.

31. McBride, "Divine Protocol," 7.

32. Welker, "What Is Creation? Rereading Genesis 1 and 2," *Theology Today* 48 (April 1991): 61.

33. This time the verb form is *qal*. According to Brown, the verb is transitive here, *Ethos*, n. 10, p. 41.

34. Ibid.

35. Ibid., 42.

36. Ibid.

37. Brett, "Earthing the Human," 75.

38. Brown, *Ethos*, 43.

39. Brett, "Earthing the Human," 77.

40. Brown, *Ethos*, 47.

41. Ibid., 51.

42. "What Is Creation?" 62.

43. Brown, *Ethos*, 45–46.

44. For a negative example, Brown points to the sin of Moses in Num 20:1–12, contending that it consists at least in part in his treatment of the rock "simply as an inert object and not as a collaborator in the creative enterprise of miracle making," which "does not give the rock its discursive due." Ibid., 97.

45. Ibid., 47.

46. Anne Gardner, "Ecojustice: A Study of Genesis 6:11–13," in *The Earth Story in Genesis*, ed. Norman C. Habel and Shirley Wurst (Sheffield, UK and Cleveland, OH: Sheffield/Pilgrim, 2000).

47. E.g., Lev. 21 and 22.

48. See also Brown, *Ethos*, 53–54. The use of nearly the same exact phrase "filled the earth with violence" in Ezek. 8:17 (also P), in which the immediate context refers to idolatrous behavior, would further substantiate that Gen. 6:11–12 has moral as well as physical connotation.

49. Gardner, "Ecojustice," 121.

50. Hermann-Josef Stripp, "'Alles Fleisch Hatte Seinen Wandel Auf Der Erde Verdorben' (Gen. 6, 12): Die Mitverantwortung Der Tierwelt an Der Sintflut Nach Der Priesterschrift," *Zeitschrift für Alttestamentliche Wissenschaft* 111, no. 2 (1999); Alexander R. Hulst, "*Kol basar* in der priesterlichen Fluterzälung," *Old Testament Studies* 12 (1958): 28–58.

51. Claus Westermann, *Genesis 1–11: A Commentary* (London: SPCK, 1984).

52. That this is the usage in all prophetic literature is a highly questionable conclusion in itself.

53. Ibid.

54. Ibid., 415.

55. 6:12–13, 17; 7:15–16, 21; 8:17; and 9:11, 15(twice)–17.

56. "Ecojustice," 121–22.

57. Brown, *Ethos*, 56.

58. Ibid., 57.

59. Ibid., 54 (emphasis mine).

60. Ibid., 56.

61. *The Yahwist's Landscape.*

62. Newsom, "Common Ground: An Ecological Reading of Genesis 2–3."

63. Ibid., 65–66.

Chapter 4. The Continuum

1. Neil Evernden, *The Natural Alien*, second ed. (Toronto and London: Toronto University Press, 1993).

2. Thomas Berry, *The Dream of the Earth*, Sierra Club paperpack ed. (San Francisco: Sierra Club, 1990), 125.

3. Robert S. Gottfried, *The Black Death: Natural and Human Disaster in Medieval Europe* (New York: Free Press, 1983), 163.

4. Europe was not the only continent affected. The plague most likely arrived from Asia, where there were similar and perhaps even greater losses, and afflicted North Africa as severely as elsewhere. The focus here is on Europe because of the connection to European modernity and scientific pursuits.

5. Gottfried, *The Black Death*, 77.

6. Ibid., 33–34.

7. Ibid., 77–78.

8. Ibid., 77–103, 129–160.

9. Ibid., 92.

10. Ibid., 91.

11. *Dream of the Earth*, 125.

12. Ibid.

13. The phrase is Gilbert Ryle's, used to interpret Descartes's thesis, in *The Concept of Mind* (London: Hutchinson, 1949).

14. *Discourse on the Method of Rightly Conducting the Reason, and Seeking the Truth in the Sciences*, Part 5.

15. Marian Stamp Dawkins, *Through Our Eyes Only? The Search for Animal Consciousness* (Oxford: Freeeman, 1993).

16. Mary Midgley, "Descartes' Prisoners," *The New Statesman*, 24 May 1999, http://www.newstatesman.co.uk/199905240041.htm; last accessed 8 February 2002.

17. Midgley, *The Ethical Primate*, 176.

18. David DeGrazia, *Taking Animals Seriously: Mental Life and Moral Status* (New York: Cambridge University Press, 1996); Dawkins, *Through Our Eyes Only?*

19. Roger Fouts and Stephen Tukel Mills, *Next of Kin: What Chimpanzees Have Taught Me About Who We Are* (New York: Morrow, 1997).

20. Ibid., 174.

21. Ibid., 180.

22. Charles Taylor, *Human Agency and Language: Philosophical Papers* (Cambridge, New York, Melbourne: Cambridge University Press, 1985), 16.

23. Frans de Waal, *Good Natured: The Origins of Right and Wrong in Humans and Other Animals* (Cambridge, MA/London: Harvard University Press, 1996), 209.

24. Robert L. Pitman and Susan J. Chivers, "Terror in Black and White," *Natural History* 107, no. 10 (1998/99): 26–29.

25. For example, Michael Ruse and Edward O. Wilson characterize our whole notion of ethics as a ruse to mask this. "Morality, or more strictly our belief in morality, is merely an adaptation put in place to further our reproductive ends. . . . Ethics is a shared illusion of the human race. If it were not so it would not work . . . [T]he way our biology enforces its ends is by making us think that there is an objective higher moral code, to which we are all subject." Michael Ruse and Edward O. Wilson, "The Evolution of Ethics," in *Religion and the Natural Sciences*, ed. James Huchingson (New York: Harcourt Brace Jovanovich, 1993), 310–11.

26. Fouts, *Next of Kin*, 5.

27. de Waal, *Good Natured*, 208.

28. John F. Haught, *God after Darwin: A Theology of Evolution* (Boulder, CO and Oxford: Westview, 2000), 1–22.

29. Verlyn Klinkenborg, "Hearing the Echo of Earthly Music: Science Again Discovers How Much We Resemble the Rest of Creation," editorial, *New York Times*, 17 January 2001: A22.

30. Haught, *God after Darwin*, 130–137; Brian Swimme and Thomas Berry, *The Universe Story: From the Primordial Flaring Forth to the Ecozoic Era—A Celebration of the Unfolding of the Cosmos*, first paperback ed. (New York: HarperCollins, 1994).

31. Haught, *God after Darwin*, 23–24.

32. I would point to the columns of Thomas Friedman in the *New York Times* in the wake of September 11 as just one example of the force of argument made on behalf of modernity yet in the public sphere.

33. "Technoscience" and "implosion" are the terms of Donna J. Haraway, *Modest Witness*.

34. Ibid.

35. Bruno Latour, *We Have Never Been Modern* (Cambridge, MA: Harvard University Press, 1993), 23.

36. Haraway, *Modest Witness*, 31.

37. Sir Winston Churchill, *Time* (New York, 12 Sept. 1960), per *The Columbia Dictionary of Quotations* (New York: Columbia University Press, 1993, 1995).

38. Kevin Kelly, *New Rules for the New Economy: Twelve Dependable Principles for Thriving in a Turbulent World* (*Wired*, 1997 [cited 20 February 2002]); available from http://www.wired.com/wired/archive/5.09/newrules_pr.html.

39. The group included Hans Moravec, author of *Robot, Mere Machine to Transcendent Mind* (New York: Oxford University Press, 1999); Ray Kurzweil, author of *The Age of Spiritual Machines: When Computers Exceed Human Intelligence* (New York: Viking, 1999); John Holland, inventor of the notion of genetic algorithms, and his former student, John Koza, who works with a method of programming that disassembles and reassembles itself modeled on DNA, along with Kevin Kelly, editor at large for *Wired*, and author of *Out of Control: The Rise of Neobiological Civilization* (Reading, MA: Addison-Wesley, 1994), a book

about the merger of biology and technology; Ralph Merkle, a leading nanotechnologist, and Bill Joy.

40. Doug Hofstadter, "Spiritual Robots: Doug Hofstadter Presentation" (*Dr. Dobb's*, 22 May 2000 [cited 20 February 2002]); available from http://technetcast.ddj.com/tnc_play_stream.html?stream_id=256.

41. Bill Joy, "Why the Future Doesn't Need Us," *Wired*, April 2000. Available from http://www.wired.com/wired/archive/8.04/joy.html.

42. http://technetcast.ddj.com/tnc_play_stream.html?stream_id=258.

43. Kevin Kelly, "Will Spiritual Robots Replace Humanity by 2100?" [Internet] (*Dr. Dobb's* Technetcast, 2000 [cited 20 February 2002]); available from http://technetcast.ddj.com/tnc_play_stream.html?stream_id=267.

44. *Modest Witness*, 109, referring to Edward Drexler et al., *Advances in Genetic Technology* (Lexington, KY: Heath, 1989).

45. "Spiritual Robots: Audience Q&A," http://technetcast.ddj.com/tnc_play_stream.html?stream_id=273.

46. *Modest Witness*, 168.

47. Ibid., 129.

Chapter 5. Reconsidering Human Moral Agency

1. Taylor, *Human Agency and Language*.

2. "From the Native's Point of View: On the Nature of Anthropological Understanding," in Richard Shweder and Robert LeVine, eds. *Culture Theory: Essays on Mind, Self and Emotion* (Cambridge: Cambridge University Press, 1984), 126.

3. Allen et al., "Prolegomena to Any Future Artificial Moral Agent," *Journal of Experimental and Theoretical Artificial Intelligence* 12 (2000): 251–61.

4. See the analysis by Kevin M. Graham, "Philosophical Theories of Justice and Agency" (Doctoral Dissertation, University of Toronto, 1996). Also Lois McNay, *Gender and Agency: Reconfiguring the Subject in Feminist and Social Theory* (Cambridge: Polity, 2000). And Michael Welker, "Is the Autonomous Person of European Modernity a Sustainable Model of Human Personhood?" in *The Human Person in Science and Theology*, ed. Niels Henrik Gregersen, Willem B. Drees, and Ulf Görman (Edinburgh and Grand Rapids: Clark/Eerdmans, 2000), 95–114.

5. Anna L. Peterson, *Being Human: Ethics, Environment, and Our Place in the World* (Berkeley, Los Angeles, London: University of California Press, 2001), 41.

6. *Taking Animals Seriously*, 67. DeGrazia argues similarly that the argument based on a principle of reciprocity—moral agents have duties only to other moral agents, and since a nonmoral agent cannot perform those duties it would be "unfair" to award it rights—breaks down under scrutiny.

7. Katie G. Cannon, *Black Womanist Ethics* (Atlanta: Scholars, 1988), 174.

8. I am not making an animal-rights argument here (or trying to undermine hard-won human rights). There is still far too much that is problematic in our construction of rights, with its tenuous relation to constricted ideas of moral agency and emphasis on individualism, that serves the purposes of domination over any other entities not accorded rights, for me to pursue the animal-rights route. See Chris J. Cuomo, *Feminism and Ecological Communities: An Ethic of Flourishing* (London and New York: Routledge, 1998), 93–95.

9. In addition to "all flesh" of the antediluvian world, there are other examples, such as the serpent (Gen. 3:14) and the goring ox (Exod. 21:28–32).

10. One only needs to think of the interpretation of Kant's principle of *Pflicht*, or duty, in the succeeding generations of his countryfolk. But closer to home, the argument of acting out of "good intentions" still holds some sway in the way whites tell the history of genocide in Canada. "Good intentions" covers a multitude of reasons for "helping" the "Indians" become as much like white folk as white folk would allow and still retain power in the land.

11. Daniel Quinn, *Ishmael*, paperback ed. (New York: Bantam, 1992/1995), 25.

12. The "we" here is very specific—First Worlders of privilege.

13. David Loy, "Religion and the Market," in *Worldviews, Religion, and the Environment: A Global Anthology,* ed. Richard C. Foltz (Belmont, CA: Thomson Wadsworth, 2003). Also available at the site of The Religious Consultation on Population, Reproductive Health and Ethics, http://www.religiousconsultation.org/loy.htm.

14. Cuomo, *Feminism and Ecological Communities: An Ethic of Flourishing*, 98.

15. Cannon, *Black Womanist Ethics*, 4.

16. Ibid., 12.

17. Ibid., 105. A common folk expression Cannon learned from her mother.

18. Ibid., quoting from Zora Neale Hurston, *Their Eyes Were Watching God* (Philadelphia: Lippincott, 1938; reprint ed., Urbana: University of Illinois Press, 1978), 119.

19. *Black Womanist Ethics*, 105, 14.

20. Ibid., 174; Cannon cites Howard Thurman to this effect: "I am saying that a man need not ever be completely and utterly a victim of his circumstances despite the fact, to be repetitive, that he may not be able to change the circumstances. The clue is in the fact that a man can give his assent to his circumstances or he can withhold it, and there are a desert and a sea between the two." Howard Thurman, "What Can We Believe In?" reprinted from *Journal of Religion and Health* 12 (April 1973): 117.

21. *Black Womanist Ethics*, 125.

22. Ibid., 126.

23. Ibid., 125.

24. Ibid., 126–27.

25. Ibid., 143.

26. Ibid., 144.

27. Ibid., 145.

28. Ibid., 144.

29. Ibid., 160–61.

30. Ibid., 162, quoting from Howard Thurman, "Mysticism and Social Change," *Eden Theological Seminary Bulletin* 4 (Spring Quarter, 1939), 27.

31. Quoted by Albert Raboteau, in his foreword to John Chryssavgis, *Beyond the Shattered Image* (Minneapolis: Light & Life, 1999).

32. Hurston, *Their Eyes Were Watching God*, 234–35, quoted by Karen Baker-Fletcher, *Sisters of Dust, Sisters of Spirit: Womanist Wordings of God and Creation* (Minneapolis: Fortress, 1998), 29.

33. *Sisters of Dust*, 34.

34. Ibid.

35. Ibid., 35.

36. Ibid., 17.

37. Ibid., 70.

38. Ibid., 39.

39. Ibid., 50–51.

40. Ibid., 58.

41. Cannon, *Black Womanist Ethics,* 164. This is part of her assessment of Thurman.

42. Ibid., 4.

43. "Mysticism and Social Change," 27; quoted by Cannon, 162.

44. Dennis H. McPherson and J. Douglas Rabb, *Indian from the Inside: A Study in Ethno-Metaphysics* (Thunder Bay: Lakehead University Press, 1993).

45. Ibid. and Rupert Ross, *Dancing with a Ghost: Exploring Indian Reality* (Markham, ON: Reed, 1992), 12–28.

46. Knockwood, *Out of the Depths,* p. 15.

47. A 1984 study for the British Columbia Ministry of Education actually used the term "autonomous" to describe Native children, in contrast to non-Native children, who are described as "dependent." Ibid., 98.

48. Gwen Bear Orechia, "The Talking Circle," in *Maliseet and Micmac: First Nations of the Maritimes,* ed. Robert M. Leavitt (Fredericton, NB: New Ireland, 1996) (emphasis mine).

49. Susanne E. Miskimmin, "Talking about C-31s: Algonquian Discourse Concerning an Amendment to the Indian Act," in *Papers of the Twenty-Eighth Algonquian Conference,* ed. David H. Pentland (Winnipeg: University of Manitoba, 1997), 255. The learning of what is expected of a person, what is good and what is unacceptable, is done through example, experience, story, and shaming, not through didactic approaches, punishment, or coercion. The sources that attest to this as a widespread practice among indigenous North Americans are numerous. Some examples: James Roger Miller, *Shingwauk's Vision: A History of Native Residential Schools* (Toronto: University of Toronto Press, 1996), 425; "The Storyteller's Lesson," Christine Saulis, *Maliseet and Micmac,* 14; Keith H. Basso, "'Stalking with Stories': Names, Places, and Moral Narratives among the Western Apache," in *On Nature: Nature, Landscape, and Natural History,* ed. Daniel Halpern (San Francisco: North Point, 1987).

50. Taiaiake Alfred, *Peace, Power, Righteousness: An Indigenous Manifesto* (Don Mills, ON: Oxford University Press, 1999), 92.

51. Ibid., 91–93.

52. It may seem contradictory that accountability is more direct and yet includes nonliving generations, except that, from the point of view of many indigenous cultures, the dead are not absent, and particularly in those indigenous cultures with a belief in reincarnation, neither are the future generations. "People do not have to look to the past to know the ancestors but are walking on their trails and reliving their lives. In this way, the trail becomes sacred." Miskimmin, "Talking about C31s," 254.

53. Battiste, "Cultural Transmission and Survival in Contemporary Micmac Society."

54. An indigenous autonomous sense of self does not preclude a sense of self that identifies with others, including the land, as a "we-self." Jace Weaver, "From I-Hermeneutics to We-Hermeneutics: Native Americans and the Post-Colonial," in *Native American Religious Identity: Unforgotten Gods*, ed. Jace Weaver (Maryknoll, NY: Orbis, 1998), 38–43; Jace Weaver, *That the People Might Live: Native American Literatures and Native American Community* (Oxford and New York: Oxford University Press, 1997); Harold Coward, "Self as Individual and Collective: Ethical Implications," in *Visions of a New Earth: Religious Perspectives on Population*, ed. Harold Coward and Daniel C. Maguire (Albany: State University of New York Pres, 2000), 43–64. For example, Georgina Tabac's testimony: "Every time the white people come to the North or come to our land and start tearing up the land," testified Georgina Tobac, "I feel as if they are cutting up our own flesh because that's the way we feel about our land. It's our flesh." As reported by Mr. Justice Thomas R. Berger, *Northern Frontier, Northern Homeland: The Report of the Mackenzie Valley Pipeline Inquiry,* vol. 1 (Ottawa: Ministry of Supply and Services Canada, 1977).

55. Rupert Ross, *Returning to the Teachings: Exploring Aboriginal Justice* (Toronto: Penguin, 1996), 270–71. Sometimes the threads of responsibility for harm as understood by Native people can be so starkly different from that of Euro-North Americans that it may be hard to comprehend. The "waves" upon which persons ride can include malevolent powers we do not discern or would regard as "superstition." See, for example, Jean-Guy A. Goulet, *Ways of Knowing: Experience, Knowledge, and Power Among the Dene Tha* (Vancouver: University of British Columbia Press, 1998), xiii–xiv.

56. Tobique Women's Group (and Janet Silman), *Enough Is Enough: Aboriginal Women Speak Out* (Toronto: Women's Press, 1987), 217.

57. Ibid., 218.

58. Ibid., 222.

59. Michael Welker, "Is the Autonomous Person . . . Sustainable?" 111.

60. Quoted by Rasmussen, *Earth Community*, 297; Dietrich Bonhoeffer, "Grundfragen einer christlichen Ethik," *Gesammelte Schriften* (Munich: Kaiser, 1966) 3:57–58; trans. Rasmussen.

Chapter 6. Doing Ethics in a Moral Habitat

1. John A. Rawls, *A Theory of Justice* (Cambridge, MA: Belknap Press of Harvard University Press, 1971). I am not taking issue in this case with Rawls's definition of justice, but with the methodology proposed for ethical discussion.

2. Donna J. Haraway, *Modest_Witness@Second_Millenium. Femaleman©_Meets_Oncomouse™* (New York: Routledge, 1997), 37.

3. Mary Midgley, *Can't We Make Moral Judgments?* (New York: St. Martin's, 1991, 1993), 82–83.

4. Ibid., 90.

5. Martha C. Nussbaum, *Love's Knowledge* (Oxford: Oxford University Press, 1990). Similarly, Colin McGinn, *Ethics, Evil and Fiction* (Oxford: Oxford University Press, 1997).

6. Greg Sarris, *Keeping Slug Woman Alive: A Holistic Approach to American Indian Texts* (Berkeley, Los Angeles, and Oxford: University of California Press, 1993).

7. David Joseph Wellman, "Sustainable Diplomacy: Ecological Realism and Muslim-Christian Dialogue in the Context of Moroccan-Spanish Relations" (Doctoral dissertation, Union Theological Seminary, 2002).

8. Marilyn J. Legge, *The Grace of Difference: A Canadian Feminist Theological Ethic* (Atlanta: Scholars, 1993).

9. Sarris, 4.

10. Cynthia D. Moe-Lobeda, *Healing a Broken World: Globalization and God* (Minneapolis: Fortress, 2002).

11. "The Rise and Fall of Natural History," *Orion* (Autumn, 2001): 18.

12. *Being Human*, 75.

13. Paul Kennedy, address to the 75th Congress of the Social Sciences and Humanities, Toronto, Ontario, 30 May 2006.

14. Aidan Davison, *Technology and the Contested Meanings of Sustainability* (Albany: State University of New York Press, 2001).

15. Kelly, *Will Spiritual Robots Replace Humanity?*

16. Berry, *Dream of the Earth*, 195.

17. Thomas Berry, *The Great Work* (New York: Random House, 1999), ix.

18. Ibid., x.

19. Ibid.

20. Ibid., 7.

21. Bill McKibben, *The Age of Missing Information* (New York/London: Penguin, 1992), 80.

22. *Technology*, 212.

Bibliography

Alfred, Taiaiake. *Peace, Power, Righteousness: An Indigenous Manifesto.* Don Mills, ON: Oxford University Press, 1999.

Anderson, Bernhard W. *Creation versus Chaos: The Reinterpretation of Mythical Symbolism in the Bible.* New York: Association, 1967.

Anderson, E. N. *Ecologies of the Heart: Emotion, Belief, and the Environment.* New York and Oxford: Oxford University Press, 1996.

Baker-Fletcher, Karen. Sisters of Dust, Sisters of Spirit: Womanist Wordings of God and Creation. Minneapolis: Fortress, 1998.

Basso, Keith H. "'Stalking with Stories': Names, Places, and Moral Narratives among the Western Apache." In *On Nature: Nature, Landscape, and Natural History,* ed. Daniel Halpern. San Francisco: North Point, 1987, 95–116.

Battiste, Marie. "Cultural Transmission and Survival in Contemporary Micmac Society." *The Indian Historian* 10, no. 4 (1977): 2–13.

———. "Mi'kmaq Socialization Patterns." In *The Mi'kmaq Anthology,* ed. Rita Joe and Lesley Choyce. Lawrencetown Beach, NS: Pottersfield, 1997, 145–61.

Berry, Thomas. *The Dream of the Earth.* Sierra Club paperpack ed. San Francisco: Sierra Club, 1990.

———. *The Great Work.* New York: Random House, 1999.

Berry, Thomas, and Thomas Clarke. *Befriending the Earth: A Theology of Reconciliation between Humans and the Earth.* Mystic, CT: Twenty-third Publications, 1991.

Bowers, C. A. *The Culture of Denial: Why the Environmental Movement Needs a Strategy for Reforming Universities and Public Schools.* Ed. David W. Orr and Harlan Wilson, *Suny Series in Environmental Public Policy.* Albany: State University of New York Press, 1997.

Brett, Mark C. "Earthing the Human in Genesis 1–3." In *The Earth Story in Genesis,* ed. Norman C. Habel and Shirley Wurst, 73–86. Sheffield, UK and Cleveland, OH: Sheffield/Pilgrim, 2000.

Brown, William P. "Divine Act and the Art of Persuasion in Genesis 1." In *History and Interpretation: Essays in Honour of John H. Hayes,* ed. M. P. Graham et al. Sheffield, UK: Journal for the Study of the Old Testament, 1993, 120–25.

————. *The Ethos of the Cosmos: The Genesis of Moral Imagination in the Bible.* Grand Rapids, MI: Eerdmans, 1999.

————. *Structure, Role, and Ideology in the Hebrew and Greek Texts of Genesis 1:1–2:3.* Ed. David L. Petersen. Vol. 132, *Sbl Dissertation Series.* Atlanta, GA: Scholars, 1993.

Callicott, J. Baird. *In Defense of the Land Ethic: Essays in Environmental Philosophy.* Albany: State University of New York Press, 1989.

Cannon, Katie G. *Black Womanist Ethics.* Atlanta: Scholars, 1988.

Coward, Harold. "Self as Individual and Collective: Ethical Implications." In *Visions of a New Earth: Religious Perspectives on Population,* ed. Harold Coward and Daniel C. Maguire. Albany: State University of New York Press, 2000, 43–64.

Cuomo, Chris J. *Feminism and Ecological Communities: An Ethic of Flourishing.* London and New York: Routledge, 1998.

Dawkins, Marian Stamp. *Through Our Eyes Only? The Search for Animal Consciousness.* Oxford: Freeeman, 1993.

de Waal, Frans. *Good Natured: The Origins of Right and Wrong in Humans and Other Animals.* Cambridge, MA/London: Harvard University Press, 1996.

DeGrazia, David. *Taking Animals Seriously: Mental Life and Moral Status.* New York: Cambridge University Press, 1996.

Evernden, Neil. *The Natural Alien.* second ed. Toronto and London: Toronto University Press, 1993.

Feit, Harvey A. "The Ethno-Ecology of the Waswanipi Cree; or How Hunters Can Manage Their Resources." In *Cultural Ecology,* ed. Bruce Cox. Toronto: Macmillan, 1978, 115–39.

Fouts, Roger, and Stephen Tukel Mills. *Next of Kin: What Chimpanzees Have Taught Me About Who We Are.* New York: Morrow, 1997.

Fretheim, Terence E. "Creator, Creature and Co-Creation in Genesis 1–2." *Word and World Suppl* 1 (1992): 11–20.

Gardner, Anne. "Ecojustice: A Study of Genesis 6:11–13." In *The Earth Story in Genesis*, ed. Norman C. Habel and Shirley Wurst. Sheffield, UK and Cleveland, OH: Sheffield/Pilgrim, 2000, 117–29.

Gottfried, Robert S. *The Black Death: Natural and Human Disaster in Medieval Europe.* New York: Free Press, 1983.

Graham, Kevin M. "Philosophical Theories of Justice and Agency." Doctoral Dissertation, University of Toronto, 1996.

Group, Tobique Women's (and Janet Silman). *Enough Is Enough: Aboriginal Women Speak Out.* Toronto: Women's Press, 1987.

Habel, Norman. "Geophany: The Earth Story in Genesis." In *The Earth Story in Genesis*, ed. Norman Habel and Shirley Wurst. Sheffield, UK and Cleveland, OH: Sheffield/Pilgrim, 2000.

Haraway, Donna J. *Modest_Witness@Second_Millenium.Femaleman©_Meets_Oncomouse™.* New York: Routledge, 1997.

Harrod, Howard L. *The Animals Came Dancing: Native American Ecology and Animal Kinship.* Tucson: University of Arizona Press, 2000.

Haught, John F. *God after Darwin: A Theology of Evolution.* Boulder, CO and Oxford: Westview, 2000.

Hiebert, Theodore. *The Yahwist's Landscape: Nature and Religion in Early Israel.* New York/Oxford: Oxford University Press, 1996.

Hofstadter, Doug. "Spiritual Robots: Doug Hofstadter Presentation" *Dr. Dobb's*, 22 May 2000 [cited 20 February 2002]. Available from http://technetcast.ddj.com/tnc_play_stream.html?stream_id=256.

Hollowell, A. Irving. "Ojibwa Culture and World View." In *Contributions to Anthropology: Selected Papers of A. Irving Hallowell.* Chicago and London: University of Chicago Press, 1976.

Johnston, Basil. *Ojibway Heritage.* Toronto: McClelland and Steward, 1984.

Joy, Bill. "Why the Future Doesn't Need Us." *Wired*, April 2000, 238–62.

Keller, Catherine. *Apocalypse Now and Then.* Boston: Beacon, 1996.

Kelly, Kevin. *New Rules for the New Economy: Twelve Dependable Principles for Thriving in a Turbulent World. Wired*, 1997 [cited 20 February 2002]. Available from http://www.wired.com/wired/archive/5.09/newrules_pr.html.

———. *Will Spiritual Robots Replace Humanity by 2100?* [Internet]. *Dr. Dobb's* Technetcast, 2000 [cited 20 February 2002]. Available from http://technetcast.ddj.com/tnc_play_stream.html?stream_id=267.

Knockwood, Isabelle. *Out of the Depths*. Lockeport, NS: Roseway, 1992.

LaBar, Martin. "A Biblical Perspective on Nonhuman Organisms: Values, Moral Considerability, and Moral Agency." In *Religion and the Environmental Crisis*, ed. E. Hargrove. Athens: University of Georgia Press, 1986, 76–93.

Legge, Marilyn J. *The Grace of Difference: A Canadian Feminist Theological Ethic*. Atlanta: Scholars, 1993.

Latour, Bruno. *We Have Never Been Modern*. Cambride, MA, 1993.

Marshall, Murdena. "Salite.A Mi'kmaq Sacred Tradition." In *The Mi'kmaq Anthology*, ed. Rita Joe and Lesley Choyce. Lawrencetown Beach, NS: Pottersfield, 1997, 48–50.

Martin, Calvin. *Keepers of the Game: Indian-Animal Relationships and the Fur Trade*. Berkeley: University of California Press, 1978.

Martin, Calvin Luther. *The Way of the Human Being*. New Haven and London: Yale University Press, 1999.

McGinn, Colin. *Ethics, Evil and Fiction*. Oxford: Oxford University Press, 1997.

McKibben, Bill. *The Age of Missing Information*. New York/London: Penguin, 1992.

McNay, Lois. *Gender and Agency: Reconfiguring the Subject in Feminist and Social Theory*. Cambridge: Polity, 2000.

McPherson, Dennis H., and J. Douglas Rabb. *Indian from the Inside: A Study in Ethno-Metaphysics*. Thunder Bay: Lakehead University Press, 1993.

Midgley, Mary. *Beast and Man: The Roots of Human Nature*. New York: Cornell University Press, 1979.

———. *Can't We Make Moral Judgments?* Ithaca: St. Martin's, 1991, 1993.

————. "Descartes' Prisoners." *The New Statesman*, 24 May 1999.

Miller, James Roger. *Shingwauk's Vision: A History of Native Residential Schools*. Toronto: University of Toronto Press, 1996.

Miskimmin, Susanne E. "Talking about C-31s: Algonquian Discourse Concerning an Amendment to the Indian Act." In *Papers of the Twenty-Eighth Algonquian Conference*, ed. David H. Pentland. Winnipeg: University of Manitoba, 1997.

Moe-Lobeda, Cynthia D. *Healing a Broken World: Globalization and God*. Minneapolis: Fortress, 2002.

Morris, Sunset Rose. "Spring Celebration." In *The Mi'kmaq Anthology*, ed. Rita Joe and Lesley Choyce. Lawrencetown Beach, NS: Pottersfield, 1997, 211–24.

Nash, Ronald J., ed. *The Evolution of Maritime Cultures on the Northeast and the Northwest Coasts of America*. Vancouver: Simon Fraser University, 1983.

Newsom, Carol A. "Common Ground: An Ecological Reading of Genesis 2–3." In *The Earth Story in Genesis*, ed. Norman C. Habel and Shirley Wurst. Sheffield, UK and Cleveland, OH: Sheffield/Pilgrim, 2000, 60–72.

Nussbaum, Martha C. *Love's Knowledge*. Oxford: Oxford University Press, 1990.

Orechia, Gwen Bear. "The Talking Circle." In *Maliseet and Micmac: First Nations of the Maritimes*, ed. Robert M. Leavitt. Fredericton, NB: New Ireland, 1996, 22.

Paul, Daniel N. *We Were Not the Savages: A Micmac Perspective on the Collision of European and Aboriginal Civilizations*. Halifax, NS: Nimbus, 1993.

Peterson, Anna L. *Being Human: Ethics, Environment, and Our Place in the World*. Berkeley, Los Angeles, London: University of California Press, 2001.

Pitman, Robert L., and Susan J. Chivers. "Terror in Black and White." *Natural History* 107, no. 10 (1998/99): 26–29.

Preston, Richard J. "The Great Community of Persons." In *Papers of the Twenty-Eighth Algonquian Conference*, ed. David H. Pentland. Winnipeg: University of Manitoba, 1997, 274–82.

Quinn, Daniel. *Ishmael*. paperback ed. New York: Bantam, 1992/1995.

Rawls, John. *A Theory of Justice*. Cambridge, MA: Belknap Press of Harvard University Press, 1971.

Rogerson, John, and Philip Davies. *The Old Testament World*. Englewood Cliffs, NJ: Prentice Hall, 1989.

Ruse, Michael, and Edward O. Wilson. "The Evolution of Ethics." In *Religion and the Natural Sciences*, ed. James Huchingson. New York: Harcourt Brace Jovanovich, 1993, 308–11.

Sarna, Nahum M. *Genesis: The Traditional Hebrew Text with New Jps Translation/Commentary, The Jps Torah Commentary*. Philadelphia, New York, Jerusalem: Jewish Publication Society, 1989.

Sarris, Greg. *Keeping Slug Woman Alive: A Holistic Approach to American Indian Texts*. Berkeley, Los Angeles, and Oxford: University of California Press, 1993.

Spencer, Daniel T. *Gay and Gaia: Ethics, Ecology, and the Erotic*. Cleveland, OH: Pilgrim, 1996.

Spretnak, Charlene. *The Resurgence of the Real*. New York: Routledge, 1999.

Stackhouse, Max L. *Ethics and the Urban Ethos: An Essay in Social Theory and Theological Reconstruction*. Boston: Beacon, 1972.

Steegmann, A. Theodore, Jr., ed. *Boreal Forest Adaptations: The Northern Algonkians*. New York: Plenum, 1983.

Stripp, Hermann-Josef. "'Alles Fleisch Hatte Seinen Wandel Auf Der Erde Verdorben' (Gen. 6, 12): Die Mitverantwortung Der Tierwelt an Der Sintflut Nach Der Priesterschrift." *Zeitschrift für Alttestamentliche Wissenschaft* 111, no. 2 (1999): 167–86.

Swimme, Brian, and Thomas Berry. *The Universe Story: From the Primordial Flaring Forth to the Ecozoic Era—A Celebration of the Unfolding of the Cosmos*. First paperback ed. New York: HarperCollins, 1994.

Tanner, Adrian. Bringing Home Animals: Religious Ideology and Mode of Production of the Mistassini Cree Hunters. St. John's, NF: Memorial University of Newfoundland, 1979.

Taylor, Charles. *Human Agency and Language: Philosophical Papers*. Cambridge, New York, Melbourne: Cambridge University Press, 1985.

Tucker, Gene M. "Rain on a Land Where No One Lives: The Hebrew Bible on the Environment." *Journal of Biblical Literature* 116, no. 1 (1997): 3–17.

Valentine, Lisa Philips. "Performing Native Identities." In *Actes Du Vingt-Cinquième Congrès Des Algonquinistes*, ed. William Cowan. Ottawa: Carleton University, 1994, 482–92.

von Rad, Gerhard. "The Theological Problem of the Doctrine of Creation." In *Creation in the Old Testament*, ed. Bernhard W. Anderson. Philadelphia and London: Fortress, and SPCK, 1966, 53–73.

Waldram, James B. "The Reification of Aboriginal Culture in Canadian Prison Spirituality Programs." In *Present Is Past: Some Uses of Tradition in Native Societies*, ed. Marie Mauze. Lanham: University Press of America, 1997, 131–43.

Weaver, Jace. "From I-Hermeneutics to We-Hermeneutics: Native Americans and the Post-Colonial." In *Native American Religious Identity: Unforgotten Gods*, ed. Jace Weaver. Maryknoll, NY: Orbis, 1998.

———. *That the People Might Live: Native American Literatures and Native American Community*. Oxford and New York: Oxford University Press, 1997.

Wellman, David Joseph. "Sustainable Diplomacy: Ecological Realism and Muslim-Christian Dialogue in the Context of Moroccan-Spanish Relations." Dissertation, Union Theological Seminary, 2003.

———. "Sustainable Diplomacy: Ecological Realism and Muslim-Christian Dialogue in the Context of Moroccan-Spanish Relations." Doctoral dissertation, Union Theological Seminary, 2002.

Westermann, Claus. *Genesis 1–11: A Commentary*. London: SPCK, 1984.

Whitehead, Ruth Holmes. *Stories from the Six Worlds: Micmac Legends*. Halifax, NS: Nimbus, 1988.

Wright, G. Ernest. *God Who Acts: Biblical Theology as Recital*. London: SCM-Canterbury, 1952.

Index

143